Stinkbugs, Stick Insects, and Stag Beetles

And 18 More of the Strangest Insects on Earth

Sally Kneidel

John Wiley & Sons, Inc.
New York • Chichester • Weinheim • Brisbane • Singapore • Toronto

Published by John Wiley & Sons, Inc.
Published simultaneously in Canada

Design and production by Navta Associates, Inc.

Library of Congress Cataloging-in-Publication Data

Kneidel, Sally Stenhouse.
  Stinkbugs, stick insects, and stag beetles : and 18 more of the strangest insects on Earth /Sally Kneidel.
    p. cm.
    Summary: Focuses on the appearance and behavior of unusual insects, such as stinkbugs, stick insects, and bombardier beetles.

nsects.] I. Title.

                                                                    99–058042

Printed in the United States of America

10 9 8 7 6 5 4 3 2 1

Contents

—69—
Part Three
Strange Insects You Probably Won't Find

—103—
Resources

—105—
Further Reading

—107—
Glossary

—111—
Index

Acknowledgments _____

I would like to thank the staff and researchers at La Selva Biological Station in Costa Rica for providing such an interesting stay for me and my family, and for answering many of our questions. While staying there I learned a lot about the tropical insects in this book, particularly the leaf-cutter ants, army ants, and botflies. Any errors in the book are purely my own and not theirs.

Introduction _____

This book is about some of my very favorite insects. They are all special to me because every one of them has something very strange about its lifestyle. Some are delightful, some are gross, some are a little scary even, but all are interesting, even astonishing. These are the kinds of bug stories that will make you want to grab someone and say, "Listen to this!" Scientists have spent years and years of careful study trying to figure out why some of these odd insects, such as the honey ants, the tsetse flies, and the leaf-cutter ants, behave the way they do. I think you'll be surprised by what they've discovered.

This book is about the oddest of the insects, but all insects are fun to study. You'll never run out of subjects, because insects are everywhere. There are over a million species of insects, more than all of the other animal species on earth put together. And they are so wildly different from each other. They have adapted to almost every environment on Earth, sometimes with evolutionary tricks that we could never have imagined.

How This Book Is Organized

The first section of the book tells you about insects that you can probably find in the United States and southern Canada. The second section describes insects that you might come across, or that you could find with some effort. In the third section are insects that are not found in the United States or are found only on the southern fringes of the United States. Some could be seen on trips out of the United States or in an insect zoo.

Each chapter covers one particular type of insect. Within each chapter you will find a description of what the insect looks like, where it lives, and why it looks so strange, or why it behaves in such an unusual way. You'll learn how its odd habits help it to survive.

Words that appear in **bold** are defined in the glossary at the back of the book. These words are in bold only the first time they appear in the text. They may or may not be defined in the text.

Check the Resources section in the back of the book for information on ordering insects, a list of insect zoos, and brief descriptions of some interesting books about insects.

What Is an Insect?

An insect is an animal that has an **exoskeleton** (a hard outer covering), six jointed legs, and three body parts: the head, **thorax**, and **abdomen**. The thorax is in the center of the body. Wings and legs are attached to the thorax.

Many types of insects have two pairs of wings. Sometimes only the second or back pair is used for flight. When this is the case, the first pair are thicker and serve as wing covers. They cover and protect the second pair when the insect is not in flight. Wing covers are called **elytra**. Beetles are one group of insects that have elytra.

Insects have two different kinds of life cycles. One type of insect life cycle involves what is called **complete metamorphosis**. An egg hatches into a **larva**, which is more or less wormlike. The larva's job is to eat and grow. It sheds its skin periodically as it grows. When it has grown as much as it's going to, it becomes a **pupa**. The pupa is inactive, and does not eat. It is enclosed in a silken **cocoon**, or perhaps in a tough skin, where it **pupates**, or transforms into an adult. This transformation is a complete **metamorphosis**, meaning a complete change in shape or form. The larva and adult usually do not look at all alike, usually live in different places, and have different diets. Most insects develop into adults by complete metamorphosis. In this book, the insects with complete metamorphosis are monarch butterflies, caddis flies, burying beetles, stag beetles, dung beetles, robber flies, botflies, tsetse flies, and all of the ants.

The other type of insect life cycle involves what is called **gradual** or **incomplete metamorphosis**. In gradual metamorphosis, an egg hatches into a **nymph**, which is not wormlike. It has legs and often looks very much like the adult, but without wings. It often has the same diet as the adult and lives in the same **habitat** (same type of plant or same type of soil, etc.). It is much more like its adult parents than a larva is. The nymph eats and grows rapidly, shedding its skin at intervals. (This is called **molting**.) The last time it sheds its skin, it emerges from the skin as an adult, usually with wings. There is no pupal or transformation stage. Because the nymph is already similar to the adult, the change does not require a long period of time inside a cocoon or other casing. In this book, the insects that undergo gradual metamorphosis are aphids, cicadas, stinkbugs, back swimmers, cockroaches, stick insects, assassin bugs, and giant water bugs.

All insects are **cold-blooded**. In fact, all animals are cold-blooded except for birds and mammals. This doesn't mean that their body temperature is always low. It just means that their body temperature is not regulated internally like that of humans and other **warm-blooded** animals. Their body temperature is determined by the temperature of the air or water or surfaces around them. So it changes throughout the day. Cold-blooded animals can regulate their body temperature somewhat by moving to a warmer or cooler spot or by sitting in the sun or moving out of the sun.

Animal Classification

Animals are grouped into categories based on their similarity to one another and on their **evolutionary** relationships. The primary, or largest, divisions of the animal **kingdom** are **phyla.** There are about 26 phyla of animals, depending on whose system of classification you use. Humans are in the phylum Chordata. This phylum includes all animals with backbones (fish, amphibians, reptiles, birds, and mammals) and a few other small, obscure animals.

Insects are in the phylum Arthropoda. This phylum includes not only insects but also spiders, crustaceans, millipedes, centipedes, and others. All phyla are divided into **classes.** Insects are in the class Insecta. Some of the other classes in the phylum Arthropoda are Arachnida (spiders and their kin), Crustacea (crustaceans), Diplopoda (millipedes), and Chilopoda (centipedes).

Each class is divided into **orders.** The following are some of the major insect orders represented in this book.

Order Coleoptera

The insect order Coleoptera includes only the beetles. It is the largest order of insects, with over 300,000 species. Beetles have hard bodies and chewing mouthparts. Adults have two pairs of wings. The outer pair are really wing covers, called elytra. These meet in a straight line down the middle of the beetle's back, and form a hard cover over the back when the beetle is not flying. The name *Coleoptera* means "sheath wings," a reference to these elytra. The inner pair of wings, used for flying, are thin and clear, and not visible at rest. In flight, the elytra are flipped up and forward, out of the way.

Beetle larvae are wormlike and segmented. They are often soft-bodied and whitish, with a distinct head and six legs just behind the head. Many are called **grubs.** After growing, the larvae turn into pupae and undergo complete metamorphosis. The diet varies.

Order Hemiptera

Hemiptera is the order of true bugs. Many different kinds of insects are commonly called bugs. But to a scientist, "bug" means an insect in the order Hemiptera, just as "beetle" means an insect in the order Coleoptera.

All hemipterans have a long, tubelike, sucking mouthpart. Those that are **predators** stick the tube into their **prey** and suck out body fluids. Those that are plant eaters use the tube to pierce plants and suck out plant juices.

A hemipteran can be identified as such by the pattern on its back, created by the wings. *Hemi* means "half" and *ptera* means "wing." Each wing is half thick and leathery, and half thin and clear. These odd wings overlap one another when closed. The half-and-half aspect of the overlapped wings creates an X on the back. The X is present on all hemipterans, but more obvious on some than others.

Hemipterans undergo gradual or incomplete metamorphosis. The young are nymphs that usually feed in the same manner as their parents, and gradually grow up without any pupal stage.

Order Hymenoptera

Most species in the order Hymenoptera are solitary and do not live in **colonies.** But the most well-known members of the order are those that live in complex social groups. These are the ants, and some of the wasps and bees. The colonies of these so-called **social insects** typically have a **queen** that lays eggs, and female **workers** that don't reproduce but take care of the queen and her young.

Some hymenopterans are wingless as adults, others have two pairs of clear wings. All of the adults have chewing mouthparts. Bees, ants, and wasps have a pinched-in middle, or waist. The sawflies and their kin do not.

Hymenopterans undergo complete metamorphosis, having a larval and a pupal stage. Many construct a nest for their eggs.

Many hymenopterans, especially the bees, are important **pollinators** of plants.

Order Lepidoptera

The order Lepidoptera includes only the butterflies and moths. The name *Lepidoptera* means "scale wings," an appropriate name because their wings, bodies, and legs are covered with colored scales. The scales come off easily if a butterfly or moth gets stuck in a spiderweb, helping it to escape.

A butterfly or moth has a soft body and a coiled strawlike mouthpart for sucking flower **nectar.** The long, wormlike larvae, called **caterpillars,** have chewing mouthparts and usually eat leaves. When full size, the caterpillars form an envelope in which to pass the pupal stage. The envelope of moths, called a cocoon, is often made of silk, while that of butterflies, called a **chrysalis,** is made of toughened skin. Inside the cocoon or chrysalis, the insects turn into adults. Because they have a pupal stage, they are said to undergo complete metamorphosis.

Order Diptera

The order Diptera includes all of the flies, and only flies. *Diptera* means "two wings." All members of the order have either two clear wings or no wings. This often sets them apart from other adult insects, which usually have four wings. Flies have large **compound eyes** and piercing, sucking, or lapping mouthparts. They feed on liquids, such as blood or nectar.

Diptera is one of the larger orders of insects. Flies are found almost everywhere. The great majority are harmless. Many are helpful to humans—as predators or parasites of other insects, or as pollinators.

Some flies give a painful bite, and some transmit diseases to humans, animals, or plants.

Members of the order Diptera undergo complete metamorphosis. The larva is usually wormlike—soft, legless, and headless. It is often called a **maggot.**

Orders are further divided into **families,** which are likewise divided into **genera,** and then **species.** Animals within each division have something in common, such as wing structure. Two animals in the same genus have more features in common than two animals in the same family. Two animals in the same order have more features in common than two animals in the same class, and so on. The only animals that are capable of interbreeding, or mating, successfully in nature, are those in the same species.

The system of classification is as follows, with the largest category first and the smallest category last:

Kingdom Animalia
  Phylum
    Class
      Order
        Family
          Genus
            Species

A Few Pointers about Catching Insects and Other Small Creatures

1. **If you flip over logs or stones looking for bugs underneath, always return the log or stone to its original position.** It's home to many creatures you may not see. The dampness and crevices underneath are just as they like it. If you don't leave the log or stone as you found it, no creatures will be there the next time you look.

2. **If you put a creature in a jar to take indoors for a day or two, always place a slightly crumpled, damp (not soggy) paper towel in the jar with it.** The paper towel will provide cracks and crevices for your creature to hide in and cling to. It will also provide moisture. Most of the dead insects that children bring to me in jars have died from lack of moisture. Your paper towel will need to be moistened a couple of times daily with a few drops of water or a spray bottle.

   A peanut butter jar is a good temporary container because it's transparent and it won't break if you drop it.

3. **Most insects and small creatures will not drink water from a dish.** Never put a dish of water in a container that you're carrying around, because it is very likely to slosh out and drown your bug. Even in a

terrarium that isn't being moved, a bug may stumble into a dish of water and drown. In nature, most tiny creatures get their moisture either from dewdrops or through their skin. The best way to water them is by spraying droplets throughout the container every day and, in some cases, keeping the soil or sand they live in damp.

4. **To make a suitable lid for a jar, use a piece of cloth held in place with a rubber band.** Plenty of air passes through cloth. (Hold your shirt over your mouth and breathe through it. You'll see.) And there are no small holes in cloth for tiny creatures to escape through.

    If your bug clings to the cloth lid, thump the cloth before removing it. If a cloth lid is not readily available, you can ask your parents to poke holes in metal or plastic jar lids.

5. **Aquatic animals do best in the same water you found them in.** You can also use bottled spring water from a grocery store, but don't use distilled water. A third option is to use tap water that has been standing uncovered for 24 to 48 hours. Chlorine will evaporate from water over a period of time. Or you can use drops from an aquarium store to dechlorinate tap water instantly. When replacing water that has evaporated, you can use water right from the tap, as long as the new water doesn't add up to more than one-fourth of the total new volume of the aquarium.

6. **Handle your pet bugs carefully.** Like all animals, insects need to be handled gently and carefully. A pinching grasp, between two fingers, is likely to injure or kill a soft-bodied insect like a caterpillar. Instead, let the caterpillar crawl on your hand or arm. A pinching grasp is okay for lifting a hard-bodied beetle or a hissing cockroach, but they too will be happier crawling on your hand or arm.

Insects You Should Not Pick Up

Insects in this book that you should not pick up include particularly assassin bugs, giant water bugs, back swimmers, and stag beetles. All can either sting or pinch. Many ants can sting or bite, too. I have not heard of robber flies biting, but since they are predators they probably can. (For more information on predators, see "Warning Coloration and Mimicry" in chapter 1.) Dung beetles may have dung on their legs, and burying beetles may have carrion, or dead meat, on them, so you may not want to pick them up with your bare hands. Be sure to wash your hands thoroughly if you do touch them.

Many aquatic insects bite, such as diving beetles and water scorpions. If you are looking for giant water bugs or back swimmers and you catch any other water insects, don't hold them. Look and then let them go.

The velvet ant is an insect you should know about. The female is not winged and looks very much like a big ant, although velvet ants are really wasps. They are about 1 inch (25 mm) long and look like they are covered with red velvet. Females give a painful sting. They wander alone over the ground, not in groups like most ants. Of course, you know not to touch the ants' relatives, the bees and wasps.

Some caterpillars have bristles that sting, particularly the saddleback caterpillar, which is green with a brown spot on its back.

Don't pick up any spider unless a grown-up tells you it's OK. Many spiders bite, and two, the black widow and the brown recluse, are dangerous.

This list doesn't cover every insect that could bite or sting, only a few of the more common ones. There are a number of useful field guides in libraries and bookstores that can help you identify other bugs. If in doubt, don't touch, just watch.

Avoid Ticks

If you plan to walk through brush and weeds looking for insects, you should find out if you live in an area where there are ticks that carry Lyme disease and/or Rocky Mountain spotted fever. You can find this out by asking your doctor, or by calling the county health department or the county agricultural extension service. The information operator or the public library can give you those numbers. If you do live in an area where ticks carry these diseases, you should wear protective clothing (long sleeves and long pants, socks and shoes, and a hat) and apply bug repellent before you go out. Change and launder your clothes when you come in, and check your body for ticks. If you find one, don't touch it. Get help from an adult in removing it. You have to be sure to get the head and jaws and not to squeeze its body while removing it.

Part I

Strange Insects You Can Probably Find

Monarch Butterflies

That's Strange!

Every spring millions of monarch butterflies, with brains the size of pin-heads, leave a grove of trees near Mexico City and fly 2,000 miles (3,200 km) to the United States. No other insects fly so far. But that's not the strangest part. The following autumn the great-grandchildren of these monarchs fly 2,000 miles (3,200 km) back, to the same area near Mexico City. These great-grandchildren have never been to Mexico before, but they fly to exactly the right place. This happens year after year. How can they fly so far? And how do they know where to go?

What They Look Like

Monarch butterflies have bright, flashy, easy-to-see colors. Their wings are orange, with black veins and black edges. The black edges are spotted with white. A monarch's wingspan is 3½ to 4 inches (9 to 10 cm). The butterfly's body is black with white spots.

Monarch caterpillars (larvae) are brightly colored, too. Their bodies are covered with black, white, and yellow bands. They grow to about 2¾ inches (7 cm) and can be found on milkweed plants.

The pupa is green with gold dots. (The casing of tough skin around a butterfly pupa is often called a chrysalis.) Shortly before the adult emerges, you can begin to see the colors of the wings through the walls of the chrysalis.

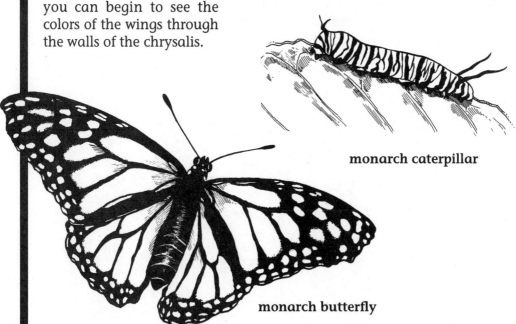

monarch caterpillar

monarch butterfly

Why Do They Do That?

Why do monarchs travel so far each year? Their journey, which is called **migration**, is a way of escaping the deadly cold of winter and the lack of food that results from the cold. In winter, most plants die or stop growing. There are no leaves or flowers for animals to eat, which means no flower nectar for butterflies. (Nectar is a sugary substance produced by flowers.) A lot of bird species migrate, but not many insect species do. Insects have another way of surviving the winter. Most insect species have one life stage that is able to get through the cold months without eating, in a resting state called **diapause**. Diapause is sort of like hibernating. For many species of insect, this **dormant** overwintering stage is the egg. Often insects lay eggs in the fall and then die, leaving the eggs behind in a sheltered place to endure winter. Praying mantises lay egg sacs that can survive the winter and hatch in the spring. Eggs don't eat anyway, so this works well. Woolly bears overwinter as larvae, or caterpillars. The caterpillar curls up under some sort of shelter on the ground and stays there in a dormant state all winter. Other insects overwinter in a pupal stage. Giant silkworm moths, for example, lie dormant in their cocoons. And some insects, such as squash bugs, overwinter as adults. Cold makes them inactive, but they may warm up and look for food on a warmish day.

Monarchs overwinter as adults. But they fly south to a much warmer place for winter, where the adults will be safe from freezing. How did monarchs first begin migrating? Monarchs eat and lay their eggs on a plant called milkweed. During the Ice Age, milkweed plants grew in Mexico, south of the United States. So monarchs lived in Mexico, too. They didn't live in the United States because there was no milkweed, and it was too cold. As the Ice Age ended, temperatures warmed and milkweed from Mexico began to move farther and farther north. Eventually milkweed spread all the way to the northern United States and some southern parts of Canada.

**monarchs
in Mexico**

As the milkweed spread north, the monarch population began to venture north in spring and summer to lay eggs on the milkweed. But the cold weather didn't suit them, so in autumn they returned to Mexico. At first the journey wasn't that far. As the monarchs continued to move farther and farther north in spring in search of milkweed, the return to Mexico got farther and farther. But they kept doing it and still do.

How do they have the energy to fly such long distances? They do fly, but they also glide on the wind, which saves energy. If the wind isn't blowing the right way, they can correct for it, or drop to the ground until the wind changes. They also make use of thermals to gain altitude. A thermal is a body of warm air moving upward in a spiral. Monarchs have been spotted by glider pilots as high as three-fourths of a mile above the ground.

How do they know where they are going? No one is sure exactly, but monarchs have some sort of inborn navigational system. Some birds, such as indigo buntings, are known to navigate by using the stars. Others, such as pigeons, have structures in their heads that can detect magnetic fields. In effect, they have a built-in compass. Bee-killer wasps have been shown to locate their nests by using visual landmarks.

The monarchs from east of the Rocky Mountains migrate to 13 sites in Mexico, all about 75 miles (120 km) west of Mexico City, and all close together. The sites are all at an elevation of 9,500 to 11,500 feet (2,850 to 3,450 m), in forested mountains. Millions of monarchs roost in fir trees for the winter months, in a dormant state. Conditions at these altitudes are cold enough to keep the monarchs inactive, but not cold enough to freeze them. The air is moist enough to keep them from losing moisture as they hang from the trees. So many monarchs cling to branches that some of the trees appear to be orange. Conditions on these mountains are perfect for their dormancy or diapause.

When March rolls around, the monarchs begin to take off. Although their trip to Mexico was leisurely, their return flight to the United States and Canada is a big rush. They're racing to the newly sprouted milkweed along the Gulf Coast of the United States, where they will lay their eggs. Many adults don't make it any farther than that. A lot of them die on the way or soon after getting there. But the eggs laid along the coast hatch, the young caterpillars get busy eating milkweed, and a month later a new generation of monarch butterflies fly north from the Gulf. After leaving the Gulf Coast for more northern areas, the monarchs go through three or four more complete generations by August. Then it's time to head for Mexico again. The adults of the final summer generation don't reproduce yet. Instead the butterflies spend a lot of time eating nectar, building up energy for the big flight. They also develop an urge to gather in groups, and an urge to move south. Then they take off for Mexico.

Warning Coloration and Mimicry

The migration of monarchs is indeed strange, but it's not the only interesting thing about monarchs. These butterflies are also an excellent example of warning coloration. Most animals on our planet are colored to blend in with their natural background—grays, browns, tans, or greens. They are **camouflaged**. Camouflage helps animals hide from predators (animals that kill and eat other animals). It also helps predators sneak up on prey without being seen.

Other animals, including the brightly colored monarchs, seem designed to attract attention. In some cases, especially among birds, bright colors attract mates. In other cases, bright colors on hidden body surfaces can startle predators when these hidden surfaces are exposed. The gray tree frog and red-eyed tree frog, for example, have bright colors on their legs that appear suddenly when they jump. This may give the frog just enough time to get away. Some insects have a brightly colored second pair of wings that can be exposed when they feel threatened. Others, such as the puss moth caterpillar or the Io moth, have bright colors on their wings that look like large eyes or a face, which they use to startle predators.

Sometimes a bright color overall can warn predators that an animal tastes bad or will cause illness. Warning colors are usually combinations of black with red, orange, yellow, or white. After a predator eats an individual with these colors, it learns to avoid others of the same species. Animals with these colors that have been shown to have **toxic** effects include orange salamanders, ladybugs, poison-dart frogs, and monarch butterflies.

Monarch caterpillars are toxic because they eat milkweed plants. Milkweed plants have poisonous chemicals in them called cardiac glycosides. Most insects and other animals are unable to eat milkweed because of these chemicals. But monarch caterpillars are immune to them. Not only are they not bothered by the chemicals, they actually store the chemicals in their bodies. This makes their bodies poisonous to

**monarch
butterfly on milkweed**

predators. Adult monarchs feed on flower nectar rather than milkweed, but the poisons stored in the caterpillar's body are still present in the adult's body. Monarch caterpillars are white, yellow, and black, and the adults are orange and black with white spots—all warning colors.

To study how warning colors work, scientists have done a lot of research with blue jays and monarchs. Blue jays eat insects, including butterflies. A scientist named Lincoln Brewer raised some monarchs on a type of milkweed that did not contain any poison. He fed the monarchs to blue jays that had been in captivity for a long time, so that they didn't know about the warning colors. The blue jays ate the monarchs, with no ill effects. Then he offered the blue jays monarchs that had been raised on toxic milkweed. The blue jays ate them as well, but soon began to ruffle their feathers and appear upset. Then they vomited. Afterward, the jays would not eat any monarchs, toxic or not. They had learned their lesson. This kind of experiment shows how birds learn to avoid insects with warning colors.

The viceroy is a butterfly that looks almost exactly like an adult monarch. The viceroy has an extra black line across each lower wing, but predators don't look this closely. A bird that has eaten a monarch and become ill will avoid an adult viceroy, too. So the viceroy is protected by **mimicry**, by imitation. An animal that benefits by looking like another animal, or model, is called a mimic. It was long thought that viceroys were not poisonous but instead just took advantage of the monarch's bad reputation. And so the viceroy was thought to be a classic example of **Batesian mimicry**, where the model (in this case, the monarch) is toxic or dangerous but the mimic is not. The mimic is only bluffing the predator. This type of mimicry is named after Henry Bates, who described it in 1861 after his travels in the Amazon valley of South America.

Now we know, however, that viceroys are not very tasty to birds either. That makes viceroys an example of **Müllerian mimicry**, where animal species that are both yucky and both have warning colors come to look like each other over time. Müllerian mimicry was first described by Fritz Müller in 1878.

Another example of Müllerian mimicry is the resemblance among many species of stinging wasps. Many have black and yellow stripes (warning colors) and make a buzzing noise (a warning sound). All stinging wasps benefit from their resemblance to each other, because if a predator, such as a toad or a bird, eats one wasp and gets stung, it is likely to avoid all similar wasps in the future.

Where They Live

Monarchs are found throughout North America, except in far northern areas. Caterpillars can be found wherever milkweed grows—in fields, in meadows, and along roadsides. Adult butterflies feed on flower nectar and may be seen in flower gardens or any areas where flowers grow. Adults may also be seen near milkweed, where they deposit their eggs.

Adults from east of the Rocky Mountains overwinter in a mountainous area west of Mexico City. They reproduce in spring along the Gulf Coast of the United States. In summer they reproduce farther northward, in the United States and Canada. They move as far north as North Dakota, southern Ontario, and Maine.

Monarchs that spend summer west of the Rocky Mountains do not migrate to Mexico. Instead they migrate to a small area of coastal California between San Francisco and Los Angeles. They perch, dormant, in trees for the winter. In California, the town of Pacific Grove celebrates the arrival of the monarchs each fall. Children dress up in monarch costumes and parade through the town. Businesses are named for the monarch. The town welcomes tourists to see the dormant butterflies clinging to trees by the millions.

The overwintering sites, in both California and Mexico, are very important to the survival of monarchs. They return to exactly the same place year after year, and are programmed to do so. If these natural areas and their trees were lost, monarchs would disappear from almost all of

migrating monarch butterflies

North America. People who care about wildlife and our environment are struggling to ensure the protection of these areas.

In a few tropical areas and in southern Florida, there are monarch populations that don't migrate.

What You Can See and Do

If you live along the Gulf Coast of the United States, you can watch for the adult monarchs just arriving from Mexico in early spring. Adults that have just made the trip may be looking tattered and may be dying. Any fresh, energetic adults that you see may be the first-generation offspring of those that arrived from Mexico. It takes only about a month to produce a new adult generation. Look in areas where milkweed grows, or among flowers.

In any part of the monarch's spring and summer range (the United States and southern Ontario), you can observe their behavior in the field. Look along roadsides and in fields of flowers for a large orange and black butterfly flying just above the flowers. Monarchs are strong fliers.

A male monarch who is ready to mate will perch in a sunny place, on the tip of a plant, watching for a female. If a large butterfly goes by, he flies out to see if it's a female monarch. He returns to his perch if it isn't. But if it is, he flies after her and bumps into her. They may mate if she is interested.

You can easily keep monarch caterpillars in captivity. Put them in a terrarium with a screen or cloth lid, or in any well-ventilated large container. They must have fresh milkweed leaves that are up off the ground so the caterpillars can grab them. Change the leaves every day, removing dead leaves and the dry pellets of caterpillar poop. Place in the container also a leafless dry multibranched stick. When a caterpillar is ready to turn into a pupa, it will leave the milkweed. It will attach itself to the stick, or the underside of some other object. Then it will shed its skin one more time and will become a pupa or chrysalis. You will be able to see the orange and black wings of the adult butterfly through the wall of the chrysalis a couple of days before it is completely mature.

When the adult comes out, its wings will be crumpled. Leave it alone, and soon its wings will straighten and dry. As soon as darkness falls, set the monarch free. As you wait for darkness, you can offer it a drink of sugar water (1 part sugar, 20 parts water). Pour a little of the sugar water on a piece of paper towel or a small sponge. If you keep the monarch butterfly, it will damage its wings flying against the cage, and will soon die. But outside it will be free to continue the migration of its ancestors.

Ants and Aphids

That's Strange!

The farmer tends her cow, gently stroking its back. When the milk lets down, the farmer carefully collects it all, making sure not to spill a drop. She leaves the cow grazing contentedly and returns to her home, to feed her hungry family. This afternoon she'll return to herd the cow to a greener pasture.

A story you've heard before? Almost. This farmer is a little different, though. When she first arrives to milk the cow, she licks it. And she collects the milk in her mouth, not in a pail. The milk is not really milk either, it's a liquid from the cow's hind end. The "cow" is an aphid and the "farmer" is an ant.

What They Look Like

An aphid is a tiny pear-shaped insect, about the size of a sesame seed. Its legs are no thicker than a hair; its eyes are specks. Many species of aphids are green or red, but they may be gray, yellow, black, or other colors.

aphid

You know what ants look like in general. Their three body parts (head, thorax, and abdomen) are clearly distinct. The thorax and abdomen are separated by a narrow waist called an abdominal stalk. All ants have an elbow joint in their antennae, which sets them apart from their narrow-waisted cousins the wasps. Ants can be brown, black, tan, red, or other colors.

Why Do They Do That?

Many species of ants and aphids all over the world interact with each other in this strange way, although not all do. Why would either ants or aphids be interested in such an arrangement? Both benefit from the relationship, just as real cows and farmers both benefit from their relationship. A human farmer takes care of a real cow—feeding it, providing shelter from harsh weather, leading it to pastures of healthier grass, protecting its young. And in return the cow eats grass and turns it into nutritious milk for us. Even though grass is abundant, it does us no good because we can't digest it. We need cows to change it into something we can eat.

The situation is very much the same for many of the ants and aphids that interact with each other. Each partnership between a particular aphid species and a particular ant species is a little different. Some ant species provide much more care than others. Some aphids need much more care.

But all of these relationships exist because aphids have the mouthparts necessary to pierce plants and feed on sap, but ants do not. Sap is nutritious and abundant, and easily eaten—if you have the right mouthparts. In a manner of speaking, the ants use the aphids to get the sap for them. Although the sap has to run through the aphid's body first, it is little altered when it comes out the aphid's hind end.

All aphids have a long, flexible beak. Although they cannot poke you with it, they can easily pierce a plant. Often they pierce one of the plant's "blood" vessels, which contain not blood but sap. A plant's sap carries **nutrients** to all the living tissues of the plant, just as blood does for us. The type of vessels that lie close to the surface in a plant stem or a leaf vein are called phloem vessels. The sap in phloem vessels is mostly sugar and water. It carries some other nutrients, but in low concentrations. To get all the nutrients it needs from the plant, an aphid must drink several times its own body weight in sap every day. So most of the sugar and water the aphid drinks passes right out its hind end. These droplets are waste, but they are not unclean. They are just sugar and water. The droplets that come from the aphid's hind end are called honeydew. A droplet may come out as often as every 10 to 20 minutes.

Aphids that are not attended by ants simply flick the droplet away with a back leg. Aphid populations can be very dense on a single plant, so honeydew droplets can build up on the leaves of the plants. The buildup can be dangerous to the aphids. Fungus may grow on the honeydew, or animals may be attracted to it. Some of these animals may eat aphids. Aphids' bodies are soft and defenseless. They are the main prey of many predatory insect species. Among them are adults and larvae of ladybugs and lacewings. (See my book *Pet Bugs* [Wiley, 1994] for more on ladybugs, lacewings, and aphids.)

Ants can protect the aphids from all of these dangers. By collecting the honeydew and taking it to their nest, ants keep the honeydew from building up on the leaves of the plant. They also protect the aphids from predators. Ants can be aggressive and can attack

ants
tending
aphids

and kill animals much bigger than themselves. Soldier ants have large powerful jaws for biting, and many can sting as well.

These are common features of ant/aphid relationships. But many ant species go far beyond this in tending their aphids. The cornfield ants of the Midwest do more for their corn-root aphids than collect honeydew and attack predators. The cornfield ants collect the aphids' eggs in the fall

cornfield ant with aphid

and protect them over the winter. When the eggs hatch in spring, the cornfield ants carry the young aphids to the roots of corn plants and other grasses, where the aphids begin to draw out sap. As the aphids reproduce over the summer, the ants continue to carry the young around to prime spots for tapping into the plants. Corn-root aphids are completely dependent on their ants. Their eggs will not survive winter unless they are in the underground nest of a cornfield ant.

═══► "Cows" in the Living Room ◄═══

Many ants of the tropical rain forest have close relationships with aphids. The acrobat ants of Central America produce an aerial nest of loose fibers, called a carton. Seeds of various plants land on the carton. They sprout and grow from the rooms and walls of the carton, until the ant nest itself becomes a garden. Inside the garden home, the acrobat ants keep their "cows" or aphids. They may keep aphid relatives, too—other small insects that suck plant juices, such as treehoppers, scale insects, and spittlebugs. The ants may get their cattle by col-

lecting eggs and bringing them into the nest. Or the aphids and their relatives may show up at the front door and be invited to stay. The aphids inside the carton are protected and have their honeydew safely removed, and ants have a handy food source in their own living room!

A close relationship between two species can be called a **symbiotic relationship**. When this relationship benefits both species, it is called more specifically a **mutualistic relationship**. The relationship between ants and aphids is mutualistic.

Where They Live

Ants live throughout the world. You can find ant/aphid relationships probably anywhere that you find aphids, certainly throughout the United States.

Aphids live on lots of different kinds of plants—trees, shrubs, weeds, and garden plants. They are often on buds or other areas of new growth, and are often on the undersides of leaves, where they are less visible. Look on tall weeds such as thistle and goldenrod in an overgrown area. If ivy grows on your house or on trees in your yard, look on the ivy leaves. Another easy place to find aphids might be in a vegetable garden, on lettuce and tomato leaves, and other leaves, stems, and buds.

What You Can See and Do

If you find aphids in a garden, on ivy, or on some weeds, you are quite likely to find a few ants wandering among them. The ants do not milk the aphids constantly, only when the aphid is ready to produce another drop. If you can watch for 15 minutes, you may see it happen. The ant will approach the aphid from behind and lick the aphid to signal that the ant is ready. The ant may also tap the aphid's back with its antennae. If the aphid is ready, it will squeeze out a drop of honeydew. Some aphids have special bristles around the hind end to hold the honeydew. The ant sucks the honeydew into its storage stomach, or **crop,** an organ for carrying food back to the ant nest. Then it can bring the fluid back up to feed the larvae, other workers, soldiers, or the queen. An ant has a separate stomach for its own meal.

You may also be able to see ants protect the aphids from a predator. Ladybugs, ladybug larvae, and lacewing larvae are very common predators of aphids. If you find some aphids and keep an eye on them over a period of several days, you may very well find one of these predators on their plant. Use a small piece of paper and a cotton swab to gently transfer the predator to an area where ants are among the aphids. How do the ants react? If you can't find one of these predators, transfer some strange ants into the ant/aphid area instead. Do the resident ants welcome the newcomers?

Could you watch ants tend aphids in captivity? Probably not. Aphids need living plants to feed on the sap. Even if you kept a live plant rooted in soil, the aphids might begin to feed again, but the ants would ignore them, at least for a while. Ants placed in a terrarium will quickly realize that the chemical trail back to their nest has been lost. After that, their only concern will be finding the trail again.

Periodical Cicadas

That's Strange!

A cicada nymph is a funny-looking insect, with bulging eyes and big hooks on the end of its front legs. The nymph of a 17-year periodical cicada lives underground for 17 years, in a small chamber, in complete darkness, all by itself. It doesn't come up for any reason until the 17 years have passed. Then suddenly this creature and all of its same-species neighbors pop out from the ground over a period of a few days to a few weeks. With no communication among them, how do they know which time is the right time?

After the periodical cicadas come out, they make up for all those years of quiet solitude. They head for the treetops and begin making a tremendous racket. Their concerts can be heard a quarter mile away!

What They Look Like

adult periodical cicada

At 1 to $2\frac{3}{8}$ inches (2.5 to 6 cm), adult cicadas are among the biggest of all insects. They have two pairs of wings. The front pair are much longer than the back pair and also much longer than the body. The wings are often transparent, and the veins in them are clearly visible. Adult cicadas have broad heads and thick, heavy bodies.

Before laying eggs, the female cicada makes a slit in the bark of a twig. Then she lays her eggs in the slit. The nymphs that hatch from the eggs are blind, white, and about $\frac{1}{16}$ inch (1.5 mm) long. They look sort of like ants except that their front legs are hooked and spiny. After leaving the egg, they fall to the ground and begin crawling around, looking for a crack in the ground or another good place to dig. They are very good at digging, with those hooked and curved front legs.

Each nymph digs down several inches and then hollows out a small chamber around itself, pressing the dug soil into the walls. The nymph is likely to be under a tree, because it fell from a tree after hatching. With luck, the cicada builds its chamber near a root of the tree. This root will be the nymph's only source of food for the next 17 years. The little creature punctures the root with its sharp beak and sucks the tree's sap. In the winter, when sap stops flowing, the nymph does nothing but sit alone in its dark chamber.

The nymph grows slowly, and every 2 to 3 years it outgrows the skin that covers its body. This outer skin, or exoskeleton, provides support for the body, since insects have no internal skeleton. Each time the exoskeleton becomes too tight, it must be shed. It splits down the back and the insect crawls out. The new exoskeleton is soft at first, and the insect expands its body as much as it can to stretch out the new one while it's still soft. Then the new exoskeleton hardens.

Why Do They Do That?

There are many species of cicadas. Only three species have a 17-year life span. Another three species have a 13-year life span. These six species are called periodical cicadas. All others have a 1-to-3-year life span. That is, the nymph stays underground for 1 to 3 years. A periodical cicada nymph has some sort of internal clock that tells it when 17 years have passed. By then, the nymph is much bigger, but it is still not considered an adult. It is about 1 inch (2.5 cm) long and quite thick-bodied. It has large eyes and well-developed hooks on its front legs. It crawls upward to the surface of the soil, or very near the surface. Then it waits. All around it, other cicada nymphs are doing the same thing. They're coming to the surface and waiting. What are they waiting for? The right soil temperature? A certain amount of daylight?

nymph digging out of its
underground chamber

Whatever the signal, when it comes, they all come out. Thousands of fat cicada nymphs may come out of the ground under a single tree, more or less at the same time.

After the full-grown nymph has left the ground, it walks over to a tree trunk (or a wall or post) and crawls up. It moves up several feet and then stops. The sharp, pointed hooks on its front legs help it to climb, and anchor it well when it stops. After a little while, the skin

cicada shedding
exoskeleton

on its back splits and the cicada slowly moves up and out of the slit, leaving the nymph exoskeleton behind. You may find these empty exoskeletons hanging on tree trunks. The insect that leaves the exoskeleton behind is now an adult. What's the difference? Its shape is the same as the nymph's, but now it has two pairs of wings. The adult holds on with its hooks as its new wings unfurl. The wings straighten as fluid is pumped into them, just as an air mattress will straighten when you pump or blow air into it. When the wings are firm and dry, the cicada takes off for a treetop. Adults feed by sucking sap, just as nymphs do. But the sap comes from twigs instead of roots.

Adult cicadas are often called locusts because so many of them take to the treetops at once. But locusts are really a type of grasshopper that travels in swarms.

Cicadas are well known for the almost deafening roar of the adults' singing. Thousands make the same sounds in perfect unison. Cicadas have unique sound-producing organs, one on each side of the abdomen. The organ is a tight membrane, like the head of a drum, but the membrane is stiffer and is curved out at rest like a bulging can lid. There are

Predator Swamping

We don't know how cicada nymphs know exactly when to leave their burrows, but we do have a good guess about why so many come out of the ground at the same time. A plump cicada nymph is a tasty meal for lots of birds, lizards, snakes, and mammals, such as raccoons, mice, and foxes. If a single nymph came out alone and struggled across the ground, it would probably be eaten. But if that individual is one of hundreds of nymphs coming out at once, it will probably survive.

Imagine someone dumping 3,000 pieces of candy on your desk once every few years and then taking them away half an hour later. You wouldn't be able to eat many pieces. Most of the candy would be "safe" from you. But if someone dropped three or four pieces of candy on your desk every day for several years, you'd probably eat all of it. When prey emerge all at once, this is called **predator swamping.** The predators are so swamped with prey that they can eat only a small fraction of it. Many small fish swim in schools for the same reason. A fish that encounters a predator while it's alone is much more likely to be eaten than a fish that is surrounded by thousands of other fish just like it.

muscles attached to the inner surface of the membrane. When these muscles contract, the membrane snaps inward, to become concave. The membrane makes a noise. Then the muscles relax and the membrane snaps outward, making a noise again. The membrane continues to snap—out, then in, then out, and so on. The cicada has a chamber of air behind the sound-making structure, as a drum, guitar, or violin does. When the sound resonates inside the chamber, the noise becomes louder. The amazing thing is the speed—the cicada can snap this membrane 300 to 400 times a second! This snapping makes a loud buzzing noise.

Other muscles can change the shape of the membrane and vary its sound. This is important because different sounds can have different meanings. In most species of cicadas, only the males make sounds. One noise is a simple buzz, to startle predators. If you pick up an adult, it is likely to make this buzz, which means "Leave me alone!" A male making a different sound, a calling song, attracts other males to join him in his tree. Over a couple of weeks, thousands of males may join in, in a single tree, singing in perfect unison like a chorus. Females are attracted by the chorus. When a female comes to the chorus, a male approaches her with a third sound, a courtship song. This song is important in convincing the female to mate with him. After mating, the female lays eggs under the bark of a twig, and the cycle starts over. Adults are short-lived, dying soon after they mate, in the same summer that they emerge from the ground.

━━ What's Your Brood? ━━

Periodical cicadas that are on the same schedule are called a brood. Broods throughout the eastern United States have official numbers. The numbering of the broods began in 1893. The 17-year cicadas that came out that year throughout the East were called Brood I. The offspring of Brood I, not to be seen as adults until 1910 (17 years later), are also called Brood I. The 17-year cicadas that came out in 1894 were called Brood II. Their offspring (also called Brood II), reappeared as adults in 1911. And so on. The 17-year periodical cicadas that emerge as adults in the summer of the year 2000 are Brood VI. They are direct descendants of the original Brood VI. Can you figure out which year Brood VI was first given a number? Start in 1893 and count. It was 1898. There are only 14 broods of 17-year cicadas. So during every period of 17 years, there are 3 years when no 17-year cicadas emerge at all. There are only five broods of 13-year cicadas. Have the missing broods become extinct? Or did they ever exist?

Where They Live

Most periodical cicadas live in the eastern United States. Cicadas occur in the West as well, but they have a 1-to-3-year life cycle.

If you live in an area where 17-year cicadas occur, you may see cicadas more than once every 17 years. You may have more than one brood in your area.

What You Can See and Do

Listen to cicadas. Can you tell them apart from crickets and grasshoppers? Crickets and grasshoppers make sounds by rubbing their legs together or against their bodies. They are likely to call at night, while cicadas are usually daytime singers. Also the noise of cicadas comes from overhead, in the treetops. Cricket and grasshopper noises usually come from near ground level (although tree crickets can be higher). Crickets and grasshoppers tend to make a constant steady hum at a low volume. A large number of cicadas together are *very* loud, so loud that they can make conversation difficult or keep drowsy people awake. Also, cicada calls seem to pulsate—*wee*-uh-*wee*-uh-*wee*-uh. Some sound more like a car that keeps cranking but won't start.

You may occasionally find a mature nymph right at the top of its burrow, or see a mature nymph perched on a tree trunk, ready to emerge from its exoskeleton. This emergence usually happens at night, when few predators are watching. You can watch the entire process, which may take a couple of hours. You can even move a mature nymph indoors and hang it on a curtain, where you can watch it crawl out and unfurl its wings. But you must not disturb the nymph while it is in the middle of the process of shedding its exoskeleton. The young adult is very soft as it comes out of its old skin, and any unnatural movement of its legs or wings will probably cause permanent damage.

Once the adult begins to walk around, and its wings look flat and dry, let it crawl up on your finger or a cloth or stick, and take it outside. Place it on a tree trunk, not the ground.

Adult cicadas are very difficult to catch, because they stay high in trees. But in late summer they begin to get old and weak. Then you may find one on the ground, having trouble flying. It may even look dead, but when you pick it up, you may find that it buzzes and moves its legs. It will gladly hang on to your shirt or hair, using the hooks on the ends of its legs. It will probably stay there as long as you leave it. Consider taking it to school for a different kind of school day. Maybe it will revive itself in class and walk to the top of your head (as close as it can get to the treetops). Or if you give it a gentle squeeze, it may give your friends a buzzy greeting.

Stinkbugs

That's Strange!

She's one tough mom. She protects both herself and her big brood of youngsters from all sorts of wild animals. If someone dangerous or unpleasant approaches, she waves them away. If that doesn't work, she puts herself in front of the children. She uses her legs to push away creatures that might pester the young ones. And if they still keep coming, look out. She's also got a chemical weapon! She'll squeeze out a nasty gob of liquid that smells so bad, and tastes so bad, the bad guys go running, no matter how big they are.

What They Look Like

Stinkbugs belong to the large insect order Hemiptera, the true bugs. This order includes many bugs other than stinkbugs. Three other hemipterans are described in this book—assassin bugs, back swimmers, and giant water bugs. All hemipterans have odd front wings. The front part of each front wing is thick and hard, but the back part is thin and more or less transparent. You can see this difference rather easily, so it is helpful in identifying bugs in this order.

southern green stinkbug

You can generally recognize a stinkbug by its shape. The back is shaped like a shield—the kind that knights used to carry. But there are other bugs that may have this shape. To be sure your bug is a stinkbug, look for a large triangular structure on the back, pointing toward the hind end. This structure forms part of the covering of the back, but is raised a little from the surrounding area. This structure is called the **scutellum.** On stinkbugs, the scutellum is not as long as the hard, thick part of the front wings. It is the relative length of the scutellum that is important in identifying a bug as a stinkbug.

Most stinkbugs are around ½ inch (12 mm) long, but they range from ¼ to ¾ inch (6 to 19 mm). The shape and size don't vary too much, but the coloring varies a lot. Some are brown, gray, or green—colors that blend in with the natural world and camouflage the stinkbug. Two very common stinkbugs that are well camouflaged are the green stinkbug and the southern green stinkbug. Both are green, with just a little orangish color around the edges. Some stinkbugs are very brightly colored, with patterns of red, orange, or yellow. The harlequin bug is a very common stinkbug that is black with bright yellow or orange markings.

harlequin bug

Thanks, Mom! (You Stinker!)

What else is strange about stinkbugs? In the first paragraph of this chapter, you read that mother stinkbugs watch over their young. But only some species of stinkbugs guard their young. A mother caring for her young is unusual among insects. We expect to see mothering behavior in **mammals** and birds, whose young would not survive without it. But elsewhere in the animal kingdom, a mother's care is the exception rather than the rule. Most reptiles, amphibians, fish, and **invertebrates** (animals without backbones) simply lay their eggs and leave them. When the young hatch, they are on their own.

A protective stinkbug mom will stand over her eggs, covering them with her body. Some continue to stand over the young after they hatch. Mom has lots of behaviors to protect junior, or her two dozen juniors. Before the eggs hatch, she may scrape the ground around them with her legs. She may also wave her antennae at an approaching insect. A mother stinkbug may turn her body so that she forms a sort of shelf between her young ones and the threat, and may buzz her wings in a noisy way. Mama stinkbug may also use her middle and back legs to kick back at whatever is bothering her young. And then she always has her stinky spray if things get desperate. (But making the stinky chemicals uses a lot of her energy, so this defense spray would be a last resort.) In at least one species, called the parent bug *(Elasmucha grisea)*, the mother continues to guard the young even when they are almost grown and wander away from her. The young ones lay down a scent trail as they move, and if they are in danger, they release an alarm scent that brings mom running along the trail to help.

The kicking, scraping, and shielding behaviors of the mother stinkbug don't really sound all that scary. Do you think these behaviors would get rid of predators? A scientist named W. G. Eberhard did an experiment to see if the presence of the mother really had any effect on the survival of her eggs. The eggs of his experimental stinkbug *(Antiteuchus tripterus)* were laid in batches of 24 or so on the underside of leaves. He removed 48 mothers from their eggs. He left several other mothers with their eggs. Then he watched the survival of all the eggs over a period of time. All of the eggs remained outdoors in their natural setting. He discovered that not a single egg survived from the 48 motherless batches! Most were taken away by ants. About half of the eggs that were guarded by their mothers survived. So mothering behavior in stinkbugs does indeed improve the young ones' chance of survival.

What happened to the eggs that died even though mom was present? Many were attacked by parasitic wasps of the family Scelionidae. These are tiny wasps that lay their eggs in or on the eggs of the stinkbug. When one of the wasp eggs hatches, the wasp larva crawls around inside the stinkbug egg and eats the developing larva. Eberhard found that the guarding behaviors of a mother stinkbug do help protect one side of her batch of eggs from the wasps. But the watchful mom never turns around! So wasps are able to sneak in from the rear and nail the eggs behind her.

parent stinkbug and nymphs

Why Do They Do That?

Smelling and tasting bad are forms of **chemical defense** in the animal kingdom. An animal is using chemical defense if it is using some type of poisonous or nasty chemical in its body to protect itself from animals that want to eat it. You probably already know about several types of chemical defense. The first chapter in this book is about the monarch, a type of butterfly that gets its bad taste from milkweed plants. The skunk, of course, is famous for its bad smell. When another animal or a person threatens the skunk, it lifts its rear end into the air and sprays a very nasty substance onto the threatening creature. The substance not only smells awful, but is irritating to the eyes and breathing passages.

A smaller animal that uses chemical defense is the millipede. If you have ever looked under a rotting log, you may have seen one. Millipedes are long and thin and have dozens of legs. When disturbed, they may roll into a tight coil. If you pick up a millipede, it will give off a foul-smelling substance that will leave your hand stinking for hours. (Centipedes look similar, but centipedes can bite. *Be sure you know the difference before picking up a millipede.* Centipedes don't coil, and they have one pair of legs per body segment. Millipedes have two pairs of legs per body segment. Most millipedes are tubular—meaning they look round in cross section. Centipedes are flat, not tubular.)

Some animals spray out a chemical defense that doesn't necessarily smell bad, but can be very irritating. Bombardier beetles spray a very toxic mixture of chemicals from their hind end when bothered. It can irritate human skin, and can seriously irritate the tender skin of a frog or toad that was thinking of eating the beetle. Some stick insects also spray out chemicals that hurt the eyes or skin of another animal. And some snakes, such as the spitting cobra, can spit poisons at enemies.

Chemical defenses are not always sprayed or oozed or spit at the enemy. Sometimes chemical defenses are injected into the enemy. Wasps and bees and some ants are very good at injecting toxic chemicals that can cause intense pain. They use a sharply pointed organ at the tip of the abdomen to deliver the jab. We call that organ a stinger. Other insects use mouthparts to deliver the poisonous or painful chemicals. Insects in the order Hemiptera, such as assassin bugs, giant water bugs, and back swimmers, are good at this. Many of them have stinging bites. So do many spiders and ants. These stinging animals often use their chemicals to kill or subdue prey, as well as for defense.

Where They Live

Stinkbugs are found throughout the United States. The harlequin bug is more common in the South.

Most stinkbugs feed on plants, but some are predators. All have long, strawlike, sucking mouthparts. When they are feeding, the sucking mouthpart or beak points down. When they are not feeding, the beak folds up against the "chest." All insects in the order Hemiptera have this type of mouthpart, including the assassin bugs, giant water bugs, and back swimmers in this book. Stinkbugs that feed on plants just suck the juices of the plants. Those that are predatory suck the juices of their prey, just as assassin bugs, giant water bugs, and back swimmers do.

Some of the plant-eating species of stinkbugs feed on a wide variety of plants; others are more picky. Harlequin bugs especially like plants in the broccoli family, including cauliflower, brussels sprouts, kale, and cabbage. But they'll also feed on potato, grape, bean, squash, sunflower, and other plants. The green stinkbug often attacks fruit trees, among other things. Berry plants are a frequent target of stinkbugs. The taste of boxed berries may be spoiled by stinkbug secretions.

If you are searching a plant for a stinkbug, look under the leaves and in areas of new and tender growth.

What You Can See and Do

If you find a stinkbug, pick it up by sliding a piece of paper or a cup under it. It may squirt some liquid on the paper or cup. Sniff the liquid. What do you think? Has the stinkbug earned its name? If you put the stinkbug in a cup, it may continue to squirt out a few drops of liquid into the cup, even if you set the cup down. The liquid may be dark, and if it is, you can easily see it. The liquid comes from small slits in the third segment of the abdomen, on the underside of the body. If you look carefully, you may be able to see the slits.

Plant-eating stinkbugs are harmless. But they may be confused with certain assassin bugs, which can bite. If you see the stinkbug with its mouthparts stuck into a plant, feeding, then you know it is a plant eater. In this case, you can feel safe picking it up. If you aren't sure, don't let it rest on your bare skin.

Bugs that suck juices from plants can't continue to feed when the stem they are on has been cut from a plant, because the juices in a cut stem soon stop flowing. So your stinkbug can't be kept in captivity for longer than overnight. If you do keep it overnight, put a damp and slightly crumpled (but not tightly wadded) paper towel in the container with it. The stinkbug can find an area in the folds of the towel that has just the right amount of moisture it needs.

Caddis Fly Larvae

That's Strange!

Imagine you're turning over stones in a stream. Under one stone you find an odd little stone tube about an inch long. You bend to look more closely and you see that the tube is not solid stone. Rather, it is made of dozens, or even hundreds, of tiny pebbles cemented together. It looks like the stones of a tubular stone chimney. This is obviously something that was put together on purpose, but by whom? A tiny stonemason? As you stare, wondering, a pair of legs appear at the end of the tube. And then four more legs. There's something alive inside the tube! Is it stuck? Trapped? As you watch, the six small legs scrabble across the surface of the big stone . . . and the little tube stays right with the legs. Whoever is inside the tube is carrying it around. But why? And where did the tube come from?

What They Look Like

Caddis flies belong to the order Trichoptera, which means "hairy wings." They are closely related to moths and butterflies. As adults, they look somewhat like moths. But caddis flies have small hairs on their wings, while moth wings are covered with flat scales.

adult caddis fly

There are thousands of species of caddis flies. The length of the adults varies from $\frac{1}{4}$ to 1 inch (5 to 25 mm). Their color may be brown, gray, or black. All adult caddis flies have four wings. When the wings are folded at rest, they form a rooflike peak over the back.

The larvae look altogether different from the adults. They have a long caterpillarlike body, with six legs, all near the head. The long body fits snugly inside a tube, or case, which may be made of stones, sand, or plant material. The larva is able to pull its head and legs inside the case for safety, as a turtle or snail withdraws into its shell. So you may see just the case, or you may see a case with legs and head poking out. The head of a caddis fly larva is not large or distinctive. The legs are more noticeable. They are pointed at the tip, and appear curved when bent, somewhat like the back legs of a crab or the legs of a spider.

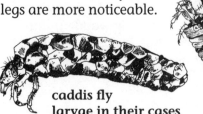

caddis fly larvae in their cases

Why Do They Do That?

The caddis fly larva builds the stone case all by itself and stays inside for protection. There are other animals that live in a protective shell of some kind—turtles, snails, fiddler crabs. But how many of them make the shell from tiny building blocks? Turtles and snails are born, or hatch, with their shells already in place. A fiddler crab gets its shell by moving into the empty shell of a seasnail, such as a whelk or conch. But a caddis fly larva has no shell when it first hatches from the egg. It has to build its shell, piece by piece. This is an amazing feat. Imagine trying to build a tiny hollow tube from grains of sand or tiny pieces of rock. How would you do it?

Most caddis fly larvae begin to build a case as soon as they hatch, and with good reason. Many bigger insects and fish would love to eat their soft caterpillarlike bodies. For a sand case, a larva begins by collecting grains of sand and gluing them together in a ring around its body. The glue comes from a **gland** on the larva's head. To make the ring longer, the larva adds grains of sand to the front end of the case. As the case grows longer, it extends farther and farther back from the larva's head until the case reaches all the way to the hind end. This back end of the case is left partially open to allow water to flow through, bringing oxygen to the larva's **gills.** The front end stays open so that the larva can come partly out to move around and to grab food. Most caddis fly larvae eat plant material.

Where They Live

Caddis flies are found throughout the United States and much of the world, wherever there are ponds and streams. They are most common in cool or northern areas, especially mountains. Adult caddis flies are **nocturnal** (active at night), and are strongly attracted to porch lights.

The females lay masses or strings of eggs in water, or drape strings of eggs across stems that are over water, so that the larvae fall into the water as they hatch. The larvae spend all of their time in water. Most of the larvae build cases that they drag around with them, but others build silken nets in the water instead. The species that build nets are predators. The nets work like spiderwebs to catch prey.

To find a caddis fly larva, look on the surfaces of underwater logs or sticks in streams or ponds. Different species prefer different spots. Some hang on to the stems of living plants, such as cattails. Others may be found among or under rocks.

In many species, the larva attaches its case to the underside of a large stone when it is ready to become a pupa. Pupation, or metamorphosis, occurs inside the case. When the time is right, the pupa leaves the case, swims to the surface, climbs out on to a plant stem or other object, and sheds its skin. A brand-new adult emerges from the skin. When its wings are ready, it flies away.

Fish can strike at and eat these newly emerged caddis flies as they cling to plant stems just above the surface. Because fish are very fond of them, many artificial fishing lures are made to look like various species of caddis flies.

What You Can See and Do

You may be able to catch a caddis fly larva in a stream or pond. Try looking under rocks and on the sides of rocks and sticks in the water. In a

Home Sweet Home

Caddis fly larvae that live in streams commonly build cases of sand or tiny pebbles. In ponds you are more likely to find larvae with cases made of plant material. Some larvae use long pieces, such as spruce needles or grass stems. The pieces are often cemented together side by side, like slats in a barrel. Or the pieces may spiral somewhat. Other caddis fly larvae build with short stems or twigs that cross the larva sideways and meet to form neat corners. The final product of these larvae is a square, tapering tube sometimes called a chimney case. Each species or type of caddis fly larva uses its own particular materials and design. The larvae of log cabin caddis flies use thin twigs that are much longer than necessary. The twigs cross and support each other, as the logs in a cabin do. But the ends of the twigs stick way out in every direction. To me, the most impressive caddis fly larva case is one that looks like a snail shell. It is constructed from sand and small stones in the form of a coiled, tapering tube. If you didn't know what it was, you would think it was a snail shell made of stone, because it has almost exactly the same shape. Caddis fly larval cases range in size from $1/4$ to 2 inches (6 to 50 mm).

CADDIS FLY LARVAE CASES

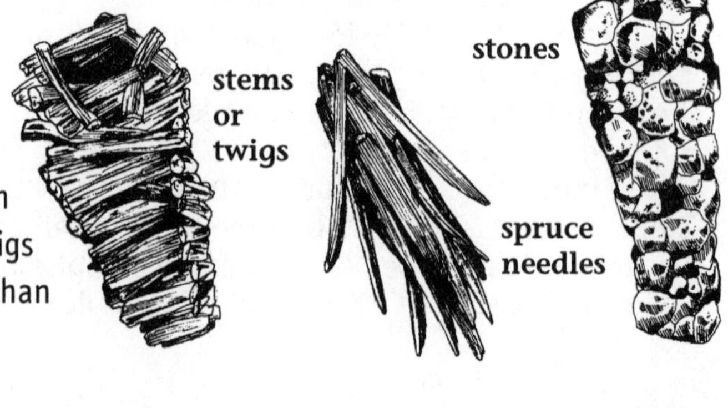

stems or twigs

stones

spruce needles

pond, you can use a net to help you look. Sweep the net through the water along the edge of the pond. As you sweep, catch a little of the plant litter from the pond floor. But do not take big scoops of sand or mud, which will weigh your net down. Mud is also hard to look through and may smother whatever you catch. Empty the contents of your net into a flat pan. If the pan is white, or clear with something white under it, you will be able to spot creatures more easily.

Look carefully on each stem and leaf. If you find the case of a caddis fly larva, you can pick it up gently between your thumb and forefinger. Place it in a clean bowl of pond or stream water. As you watch, the larva will come partway out of its case and move around, dragging the case behind. Move a toothpick or a small twig around in front of the larva. Does it react? How does it respond if you gently touch one of its legs with the toothpick?

You can't pull the larva out of the case without hurting it, so don't try. It has two hooks or humps that anchor it into its case.

To keep a caddis fly larva for more than a few minutes, you will need an aquarium with a pump or some device to put air into the water. You can get one anyplace that sells aquarium supplies.

Caddis fly larvae will repair or build cases with odd materials if you supply them. Place in the aquarium any sort of material that is the same general size and shape as the objects already in the larva's case. If its case is of small stones, offer small colored beads or small pieces of colored stone. Metal filings can be offered, too. To encourage the larva to use the new materials, you can carefully remove a few stones from its case with tweezers. The larva may use the new objects to repair the gaps. It may even be willing to build an entirely new case using the new materials you've provided.

A caddis fly larva that builds with plant materials probably won't use stonelike objects. You'll have to provide something similar in shape, size, and texture to the materials it used originally.

To feed a tube-dwelling larva, offer a variety of plant material. There are many families of caddis flies and each has a somewhat different diet. Put in your aquarium some decaying plant matter (bits of dead leaves and so on) from the pond or stream floor where you found the larva. A couple of spoonfuls is enough. Try also a small amount of algae and some living aquatic plants, or pond weeds, which are likely to have algae growing on them.

If you don't observe your caddis fly larva eating after a few days, you should probably let it go in the same spot where you found it. Don't put a stream-dwelling larva into a pond, or it will die from lack of oxygen. Stream-dwelling larvae are adapted to stream water, which has more oxygen in it than pond water. Streams pick up oxygen as the water burbles and splashes over rocks.

Burying Beetles

That's Strange!

You have just been born into this world. You stretch and wiggle, wondering where the rest of the family are. It's very dark in your little room, although this doesn't seem odd to you, since you have never seen light. But the smell—whoa! It's so strong that you can hardly think of anything else. Does the whole world smell like this? You hear strange scraping noises, and the noise seems to beckon you. You know you must go to it, and so you do.

You move along slowly on your belly, getting closer and closer to the noise and the smell. Then suddenly two giant beetles loom over your head! This must be Mom and Dad. And they've made you a snack—how kind of them! It's a big gray puddle of something rotten. They've chewed up a dead mouse and regurgitated it, or thrown it up, just for you and your siblings. You hesitate, although your siblings are diving into the gray goo with relish. Then you think about how much time it must have taken Mom and Dad to chew up the mouse carcass and digest it for you, not to mention throwing it back up.

Oh, what the heck, you think. *It could be tasty.* With a hopeful attitude, you open your mouth and lurch forward.

What They Look Like

If you were this larva, your parents would be adult beetles of the genus *Nicrophorus*. They are sometimes called burying beetles because they bury small dead animals. Another common name is sexton beetle. In previous times, a sexton was a person who buried the dead. These beetles are also called carrion beetles because they feed on carrion, which is dead meat. They share the name "carrion beetle" with beetles in the genus *Silpha,* which feed on carrion but do not bury it. *Silpha* and *Nicrophorus* beetles are cousins—both belong to the family Silphidae.

There are many species of burying beetles in the genus *Nicrophorus*. Most are black with bright orange or red markings. They range in length from $\frac{1}{16}$ to $1\frac{1}{2}$ inches (1.5 to 37 mm). Most are over $\frac{3}{8}$ inch (10 mm). They and their cousins of the genus *Silpha* are the largest beetles found on carrion, but not the only ones.

The wing covers, or elytra, of beetles in the family Silphidae are likely to leave one or more abdominal segments near the hind end uncovered. This helps to distinguish them from other beetles.

adult burying beetle

Why Do They Do That?

Why would a beetle feed its youngsters regurgitated rotten meat when there are so many other nice things to eat on this planet? Actually, a great many animals eat carrion. Animals aren't very squeamish about such things. Dead meat contains many nutrients, and there is a steady supply of it since all animals die sooner or later. Creatures that feed on dead animals or dead plant matter are called **scavengers.**

At least four different families of beetles feed on dead animals. And then there are mammals like coyotes, birds such as vultures and crows, fly larvae (maggots), and many, many more animals that are scavengers. All of these animals provide a valuable service when they break down the bodies of dead animals. They return nutrients to the soil, and they clear the land of dead animals. They are important recyclers.

The burying beetles of the genus *Nicrophorus* are the only beetles that actually bury the dead animal. They may look like they're being tidy, but they're actually doing it to save the meat for their young. Burying the carcass keeps other animals away from it, especially flies. If flies lay too many eggs on it, the fly larvae that hatch from the eggs will eat the whole carcass in no time. When the carcass is underground, flies can't get to it. And bigger animals can't steal it.

But how can an inch-long beetle bury an animal that may weigh a quarter pound? First of all, the beetle that finds the carcass has to find a mate to help it. While waiting for a member of the opposite sex to show up, the beetle guards the carcass, trying to keep competitors away. When a willing mate shows up, they begin working together. They begin by finding a good place to dig. If the soil under the carcass is too hard or rocky, then the beetles lie on their backs and push the dead mouse (or other small animal) with their legs. When the beetles have moved the

two burying beetles with carcass

carcass to an area of soil that is soft enough, they begin to crawl under the carcass and dig. They may adjust the position of the carcass with their legs. As they dig, a ridge of dirt begins to appear around the entire dead animal and the carcass begins to sink below ground level.

The beetles continue to dig, shaping a chamber under the carcass. It sinks lower and lower, until finally the soil around the hole falls slowly over the dead body, covering it up. If any roots are in the way, the burying beetles cut them with their jaws. When the digging is done, the carcass is enclosed in a little room about 1 inch (24 mm) under the surface of the ground.

Then the pair of beetles go to work on the carcass itself. They remove all of the fur and roll the carcass into a ball. They put secretions on the carcass that keep it from rotting too fast. At some time during this period, the two beetles mate and the female digs a little passageway next to the chamber. There she lays her eggs. Then the parents wait.

When the eggs hatch, the adults make a sound by rubbing their elytra or wing covers on their abdomens. This attracts the larvae to the main chamber. There, they feed upon some of the carcass that the parents have eaten, partially digested, then regurgitated for them. After a while, the larvae are able to feed on the carcass directly.

After the burying beetle larvae are grown and ready to become pupae, their parents build an exit tunnel and then leave the burrow. During the month-long period of pupation, the young will not eat and will not need their parents' care. When pupation is over and the young are new adults, they too will use the exit tunnel and leave the burrow. They are ready to search out a fresh carcass and begin the cycle over again.

Where They Live

Burying beetles are found throughout North America, in woods and pastures.

What You Can See and Do

If you find a small dead mammal or bird outside, leave it alone and observe it to see what happens. Don't touch it, but check it every day. When burying beetles are under a carcass digging, the carcass may look like it is moving and rippling on its own. You may not be able to see the beetles, but maybe you will get to see them pushing the carcass with their legs.

At the carcass you may also see carrion beetles of the genus *Silpha*, which are dull black and may have yellow or orange on their **pronotum** (the top covering of the **prothorax**, behind the head). They are broader and flatter than beetles of the genus *Nicrophorus*. Beetles of the genus *Silpha* lay eggs on or near dead animals, so that their larvae can feed on the meat. But they don't bury the carcass.

While you are looking, watch for rove beetles, too. These are beetles in the family Staphylinidae. Many species of rove beetles eat maggots on carcasses, and they may visit carcasses in large numbers. Rove beetles vary in size, but most species are smaller than burying beetles. Some are tiny. Their wing covers (elytra) are short, leaving the abdomen uncovered. Many species run around with the abdomen curled up in the air. The raised tip may look like a stinger, but it's not.

rove beetle

◄ Thanks, Mite! ►

A small carcass that burying beetles bury will almost certainly have some fly eggs on it. Flies find carcasses and lay eggs on them very quickly. And the burying process takes a while. What will happen to the maggots that hatch from the fly eggs? Won't they gobble up the whole carcass?

The parent burying beetles eat a lot of the maggots, but they also have some helpers. Beetles of the genus *Nicrophorus* have some tiny spiderlike creatures called mites that live on their bodies. When the beetles bury a carcass, the mites leave the beetles' bodies temporarily to eat the fly eggs and hatching fly larvae on the carcass. This works out well for the beetles and their young. The carcass will not be consumed by the fly larvae or maggots, so there'll be plenty of meat left for the beetle larvae.

Back Swimmers

That's Strange!

A small moth has fallen onto a pond by accident. It's stuck on the surface of the water, struggling, unable to break free. Its struggles grow weaker. It seems that the poor moth is stuck for good. There's no one around to help it out. But wait! Is that a tiny rescue boat on the way? It looks like a little rowboat, a long thin one. It has a sharp keel along the bottom as rowboats do, to keep them moving in a straight line. And look how fast those two oars are going! Whoever is paddling must really be worried about that moth.

The boat arrives, but where is the oarsman? How odd! Only four folded legs on top of the boat, and the two oars...and a sharp beak...and two big red eyes! It's a bug! And an ugly one! The bug grabs the moth, and plunges its beak into the moth's body.

What They Look Like

Not only does the back swimmer have oarlike back legs, but the whole bug is upside down! What if you saw a cat scooting across the road on its back with its legs straight up in the air? Or a bird flying upside down? This bug looks just as odd, swimming around on its back.

There are many species of back swimmers, but all are more or less the same shape—an elongate oval, like a torpedo. All have two very long back legs, which stick out from the side of the body like oars on a rowboat. These back legs are twice as long as the four front legs, which usually lie folded on the "chest" area. You can see all the legs easily from above, since the bug is lying on its back.

Back swimmers are easily recognized by their upside-down position, whether swimming or floating at the surface. When floating and resting, they hang just under the surface. The tip of the abdomen touches the surface, while the head hangs lower, so the body is at an angle. The long back legs point downward in this resting position, like oars.

One bug that you might confuse with the back swimmer is the water boatman, which also has oarlike back legs and is about the same size and shape. But water boatmen never swim upside down. Diving beetles and water scavenger beetles are also a similar size and can hang from the

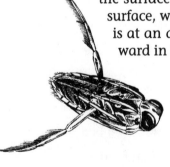

adult back swimmer, using its back legs to swim

surface, but they don't hang or swim belly-up either. Whirligig beetles swim around on the surface, but right side up and in large groups.

Back swimmers are usually dark on the belly side and light colored on the back side. One of the most common species of back swimmers is *Notonecta undulata.* Its belly side is black. Its wings are white with red areas, and dark at the tips.

All back swimmers are in the insect order Hemiptera. Other hemipterans in this book are stinkbugs, assassin bugs, and giant water bugs. All hemipterans have long, stiff, strawlike mouthparts for sucking up dinner. The sucking mouthpart or beak flips down for eating. When not in use, the beak is held flat against the "throat" or "chest" area.

Why Do They Do That?

You might feel a little uncomfortable swimming on your back because you can't see where you're going in this position. But the back swimmer doesn't have this problem. Its big eyes are at the tip end of its body and can see ahead quite well when the insect is upside down.

Being upside down may help back swimmers catch prey. They often eat insects that have fallen onto the surface of a pond. If a back swimmer is clinging to an underwater plant and sees a bug on the surface, it may just let go and drift toward the surface. Because back swimmers are always belly-up, the belly side with legs for grabbing reaches the prey first. This floating-upward-to-dinner strategy wouldn't work if the back swimmers weren't upside down.

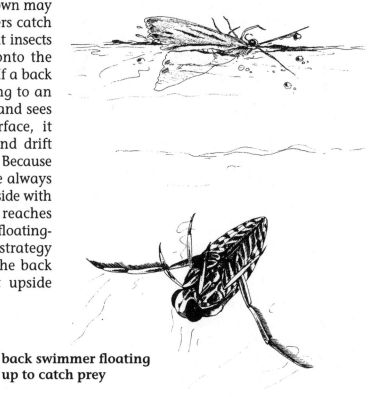

**back swimmer floating
up to catch prey**

Built-in Scuba Gear

Like whales, porpoises, seals, and manatees, back swimmers are aquatic animals that breathe air. But how? Whales and porpoises come to the surface and breathe through a hole on the back called a blowhole. Seals stick their nostrils above the surface to breathe. Back swimmers must also come to the surface to get air, but they carry the air underwater with them, trapped on the outside of their bodies. They collect the air while resting at the surface. The back swimmer sticks the tip of its abdomen above the surface a little, and special channels or grooves on the surface of the abdomen fill with air. Rows of hairs hold the air in place, so it stays in the channels when the back swimmer swims below the surface. Air is also held under the wings and in spaces between the head and thorax. When the stored air is needed for breathing, it is taken inside the body through small openings in the sides of the abdomen called **spiracles.**

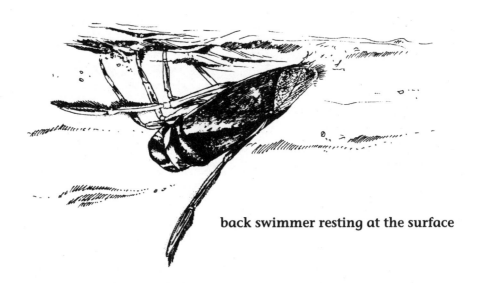

back swimmer resting at the surface

These predatory bugs, the back swimmers, eat lots of different kinds of insects—beetle larvae and other insects—as well as tadpoles and small fish. The back swimmer grabs its prey with its four front legs and then pokes its stiff beak into the prey. Substances are injected into the prey that turn the inside organs of the prey into liquid mush. Then the back swimmer slurps up these juices through its beak.

Some animals that float at the surface have light-colored bellies. This means that if a fish looks up at them, they blend in with the sky and are hard to see. Likewise, a gray or brown back on a floating creature is hard for a bird to spot from overhead. The gray or brown blends in with the water or the pond floor.

Back swimmers **evolved** or descended from long-ago insects that swam and floated right side up, but back swimmers don't have light bellies and dark backs. If they did they would be very vulnerable to fish and birds. So the light and dark colorations have flip-flopped in back swimmers. The change happened over a long period of time. As ancestral back swimmers began swimming on their backs, those with somewhat lighter backs and darker bellies survived better. Fewer of them were eaten by fish, so they survived long enough to mate and lay eggs. Their offspring inherited the characteristics of lighter backs and darker bellies. Over thousands of generations, their backs became lighter and lighter and their bellies became darker and darker. Those with the darkest backs and lightest bellies in each generation were the ones most likely to be eaten and leave no offspring behind. That's how evolution works.

Where They Live

Back swimmers are very common and are found throughout the United States. Look in ponds, small lakes, slow-moving shallow streams, and backed-up streams or water edges where water is still. Remember that back swimmers at rest hang just *under* the surface, not *on* the surface.

If you are using a net, sweep through underwater vegetation and open water in addition to looking on the surface.

What You Can See and Do

Back swimmers are hard to catch. They are very wary and will dive deep if disturbed in any way. They have excellent eyesight and can also sense vibrations in the water, so it is not easy to sneak up on them. But if you happen to catch one, you can keep it briefly in an aquarium. Put a lid on the aquarium, because a back swimmer can fly. It will eat small insects that you drop on the surface, such as moths or caterpillars.

A back swimmer can sting you with its beak, so don't touch or hold one with your bare hands. If you need to move one, use a net or cup.

You can play a trick on the tricky back swimmer if you have one in an aquarium. Put the aquarium on a glass table in a dark room, or place it so that one-quarter (no more!) of the aquarium hangs over the edge of a table. Then shine a flashlight into the aquarium from below. How do you think this will affect the back swimmer? The bug will flip over and swim with its back up and belly down! The back swimmer is confused by having light below and darkness above. Apparently its instinct is to keep its belly to the light at all times. This little experiment shows that the back swimmer orients itself by light rather than by gravity or by the relative position of air and water.

Part II

Strange Insects You Might Find

Hover Flies

That's Strange!

Puddles of water that collect on manure piles don't seem like very interesting places. Foul perhaps, but not very interesting. If you looked over the surface of such a puddle, you wouldn't see very much at all. If, however, you were able to see under the surface of a puddle like this, you might see something very interesting indeed. Amazing, even. A tiny submarine, less than an inch long! It's shaped more or less like a cucumber, with faintly visible rings around its body. On one end of the submarine is a long periscope, reaching several inches to the surface of the puddle! Is this a periscope for viewing the outside world? No, not really. What looks like a periscope is really a breathing tube, because this little submarine is really a type of fly larva. It's called a rat-tailed maggot, named for its odd breathing tube. It's the offspring of a hover fly.

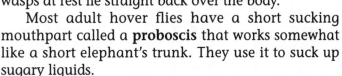

What They Look Like Most adult hover flies look like bees or wasps. They are similar in body shape and size, and they have clear wings. Many also have black and yellow patterns on their abdomens, as bees and wasps do. Their bodies may be

drone fly

covered with short fuzzy hairs, like bees. Others are hairless, like wasps. Some species of hover flies have other details that seem to mimic bees or wasps, such as front legs held to look like long wasp antennae.

But hover flies are not identical to bees and wasps. There are differences. All flies, members of the order Diptera, have only two wings. Hover flies are dipterans, so this is true for them, too. But bees and wasps have four wings, as most insects do.

The wings are held differently, too. At rest, the wings of hover flies stick out to the sides at an angle. The wings of bees and wasps at rest lie straight back over the body.

Most adult hover flies have a short sucking mouthpart called a **proboscis** that works somewhat like a short elephant's trunk. They use it to suck up sugary liquids.

The larvae of hover flies live in lots of different places. They have a variety of adaptations for different lifestyles, such as the breathing tube of the rat-tailed maggot. Most have more or less wormlike bodies. Some look very much like the larvae or maggots of more familiar flies, such as houseflies or blowflies.

rat-tailed maggot

Why Do they Do That?

Hover fly larvae have a wide range of diets and habitats, some stranger than others. The rat-tailed maggot has perhaps the strangest, living in its manure puddle with its long periscope or breathing tube. The open end of the tube is kept afloat at the surface of the puddle by a ring of hairs sticking out like the spokes of a bicycle wheel. While it breathes through the tube, the rat-tailed maggot feeds on nutrients from the manure. If the puddle changes in depth, the tube changes in length, like a collapsible telescope. These larvae can live on wet carrion as well, or in primitive outdoor toilets in the ground. When the rat-tailed maggots are fully grown, they pupate in soil for about 10 days.

Rat-tailed maggots are not the only hover fly larvae that live in water or decaying matter. But many others live elsewhere. Some live in soil or feed on living plants. Many hover fly larvae are predatory—they eat live aphids. Aphids are tiny pear-shaped insects that live on plants and feed on plant juices. (See page 10.) Predatory hover fly larvae crawl among the aphids, swinging the front ends of their sluglike bodies from side to side. When a larva locates an aphid, it pierces the aphid with its mouth hooks, holds the aphid up in the air, and sucks it dry. If you see an adult hover fly briefly touch a group of aphids, it has probably laid an egg

hover flies and bees in a flower garden

among them. The egg will soon hatch, with aphids for dinner right at hand.

Most adult hover flies feed on flower nectar, just as bees and wasps do. They fly from flower to flower sucking up the sugary liquid that the flower provides for them. If you see a number of bees and wasps flying around in a flower garden, it's likely that some of them are really hover flies. How can you tell the difference? Hover flies are much better at hovering. They are probably the most skilled of all the insect fliers. Bees and wasps bob up and down while hovering, but hover flies can remain in exactly the same place. This is much harder than it sounds, because the air itself is not still. There are always sideways currents of air, updrafts, or air currents around flowers. Yet the hover fly is able to keep its position steady. Hover flies have huge compound eyes that probably help with this skill. Compound eyes are made up of many separate visual units. They are very sensitive to movement. The hover fly can use its eye to fix its position in relation to nearby objects. When a breeze begins to shift the hover fly's position, then its wings can quickly make the necessary correction. This hovering skill is useful in collecting nectar from small spaces in flowers.

Hover flies put their outstanding flight abilities to use in other ways, too. Males of some species hover to attract mates. Some patrol an area of flowers, flying along the same route over and over. If a patrolling male sees a female, he'll fly after her and try to mate with her. Males also hover sometimes around areas where females lay eggs. They are sometimes able to intercept a female as she arrives.

Why do hover flies look like bees and wasps? The resemblance probably helps them to avoid being eaten. Birds and other predators that have learned to avoid the black and yellow warning colors of bees and wasps are likely to avoid hover flies as well. The fact that hover flies hang out around flowers helps the disguise.

Since hover flies are actually stingless and harmless, this disguise is a form of Batesian mimicry. You may remember from chapter 1 that in Batesian mimicry, the model is dangerous or painful, but the mimic is harmless. In Müllerian mimicry, both the model and the mimic are harmful.

Some hover flies take advantage of the resemblance to bees and wasps in another way. Some lay their eggs in the nests of bees and wasps. These stinging insects are usually very quick to defend their nests against invaders, but hover flies are accepted and so are the larvae that hatch from the eggs. Maybe that's because of the mimicry. Or maybe it's because the hover fly larvae do no harm. They eat only from the bees' or wasps' trash pile—dead bee bodies and other edible rubbish.

Important Deliveries

The male parts of a flower produce a yellow powdery substance called **pollen**. If you've ever touched the center of a flower, you've probably gotten pollen on your finger. Grains of pollen contain the male reproductive cells, or sperm cells, of a plant.

The female part of a flower contains egg cells that must be fertilized by pollen from another flower. When they are fertilized, the egg cells will develop into seeds. Without pollen, there will be no seeds. And, of course, new plants grow from seeds. So pollen is very important. But how does pollen get from one flower to another?

Insects and other animals carry pollen from flower to flower. They visit flowers in order to drink nectar. While looking for the drops of sweet liquid, they accidentally brush up against the pollen on the male parts of the flower. The pollen sticks to their bodies. When they fly to another flower, some of the pollen rubs off on the female surface of that flower. And that's all the help the plant needs. From there the sperm cells in the pollen can get to the egg cells so that seeds will grow.

Creatures that carry pollen from flower to flower are called pollinators. They include bees, hover flies, wasps, hummingbirds, butterflies, moths, bats, and even ants. Hover flies are very important pollinators, second only to bees. Many of the vegetables and fruits we eat are from plants that depend on bees and hover flies for the delivery of pollen, called **pollination**. So bees and hover flies are very important to our diets!

One of the most common hover flies pollinating our farms and gardens is called the drone fly. It is the adult form of the rat-tailed maggot. It is named for a male honeybee, a drone, because it looks very much like a common honeybee. Anyone would easily assume that it *is* a honeybee. Its resemblance is a disadvantage in one way. There is a type of robber fly called a bee assassin that catches and eats honeybees. The drone fly looks so much like a honeybee that bee assassins catch and eat both of them! (See page 60 for more about robber flies.)

Where They Live

Hover flies live throughout North America.

What You Can See and Do

You can observe adult hover flies in the same places where you see bees and wasps feeding on flowers. Watch the hovering behavior of any bee-like insect. If the insect bobs up and down while hovering, it's probably not a hover fly. But hover flies may rock slightly from side to side in place as they make flight adjustments.

When a beelike insect lands, look at its wings to see if they stick out like a hover fly's, or if they lie straight back like a bee's or wasp's.

You may be able to find some aphid-eating hover fly larvae. Look for them on the flower plants where the adult hover flies are feeding, or on any plants that have aphids. Aphids feed on a wide variety of plants, including tomato leaves, lettuce, and many other garden vegetables. Aphids tend to cluster in areas of new growth and on the underside of leaves. The hover fly larvae that feed on aphids are sometimes green. They generally lift the aphid in the air when feeding. The larvae's body shape is very similar to that of garbage can maggots—no eyes, no distinct head or legs, and tapered toward the mouth end. Ladybug larvae and lacewing larvae also feed on aphids and have a similar body form and size. But lacewing larvae have six legs and two very long sharp mouthparts that come together like ice tongs. Ladybug larvae also have six legs and have bands or spots of color, as well as spines or bristles. (See another of my books, *Pet Bugs* [Wiley,1994], for illustrations of lacewing and ladybug larvae.)

Don't try to catch a hover fly. You may make a mistake and get stung by a bee or a wasp.

hover fly larva eating aphid

Stick Insects

That's Strange!

You're visiting a tourist lodge in a rain forest. Tomorrow is your last day. You've seen most of the animals you wanted to see, except for the sloth. You really wanted to see that sloth, and now the guide at the lodge tells you that he just saw it. "If you go now, it will still be there. Ten minutes down the trail," he says. Your mom told you not to go down the trails alone, but she's napping. What the heck—you'll be right back.

As you start down the trail, you notice how tall the trees really are. You can't even see the tops of them, or the sky. The vegetation around you is lush, thick, and steamy. The hot and humid air is almost too heavy to breathe. Water drips from the leaves. The trail is so muddy from the afternoon rain that walking is difficult. As you look for dry places to step, you see a snake beside the trail. Your heart leaps! But it's a small one, so you keep going. Then a peccary, a wild pig, runs across the trail ahead of you. Now you *are* a little nervous. Was it peccaries they told you to stay away from, or was that something else? You can't remember. Maybe you should turn back.

You realize it's beginning to get dark, so you decide to skip the sloth and turn around. You start back, walking fast and faster. Too fast—you slip in the mud! You grab a hefty twig on a nearby shrub for support. It comes away easily in your hand, and you land on your rear in the mud—still clutching the 10-inch twig. Twig?! You see now that the twig has *legs* and they're grabbing at your arm! *Aaiiee!* Mom was right! You shouldn't have come alone!

What They Look Like

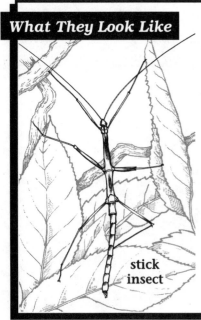

stick insect

Stick insects look remarkably like long thin twigs. They are also called walking sticks, or sticks for short. On a tree or shrub, they are very hard to spot. The color is usually green or brown, and most are wingless. When wings are present, they too blend in with the twiglike appearance—at least when they are closed. Even the small head looks like part of a twig. The antennae are long and very thin, as thin as thread.

There are over 2,500 species of stick insects worldwide. In the United States most adult stick insects are at least 2 inches (5 cm) long. Some reach 6 inches (15 cm) or longer. In tropical parts of the world, stick insects may grow as long as 10 to 12 inches (25 to 30 cm), not counting legs or antennae.

Why Do They Look Like That?

Why would an insect look like a twig? Twigginess is a form of camouflage that protects the insects from predators. Birds, lizards, and other animals that eat insects simply don't see the stick insect because it looks so much like part of a tree or shrub. Not many animals will eat twigs, because they are tough to chew.

Camouflage is very common in the animal kingdom. Most wild animals are camouflaged in some way, often just by color. An animal that is green or brown usually blends in with the background colors of the natural world. Stripes are usually camouflage, too. The stripes of tigers help them to blend in with the shadows in tall grasses where they prowl.

Some animals can be brightly colored and still be camouflaged. Flower spiders are often yellow or pink, because they hide on yellow or pink flowers, waiting for flying insects to fly in.

Some stick insects have other ways to avoid being eaten. Those with wings may flash their wings at a predator to scare it. Winged stick insects have two pairs of wings. The outer pair are camouflaged, but the inner pair are brightly colored and can be startling. Stick insects, and many other kinds of insects, will drop to the ground if disturbed. Aphids do this, too. Then the predator that disturbed them loses sight of them. If a stick insect is grabbed by a leg, the leg comes off the body easily. Then the predator is left with the leg, while the stick insect walks away. If it is a young stick, and still growing, it will get a new leg the next time it sheds its exoskelton. Crickets also drop legs easily.

stick
insect
with
wings

A few stick insects take more drastic measures to protect themselves. Some have spines on the third pair of legs that can poke and injure a predator. A very few species of stick insects can spew a toxic liquid at predators. Most, however, rely on camouflage colors and their twiggy appearance to protect them.

Stick insects have another outstanding talent that has nothing to do with protection from predators. Some populations of sticks are able to carry on for years and years, producing new generations of young sticks, without any males at all. All girl populations! But why? Stick insects move slowly and may not cover much ground. They may not encounter individuals of the same species very often. This could be a problem if you had to find a mate in order to reproduce. So if you could reproduce without a male, that would be an advantage.

The production of live young from unfertilized eggs is called

parthenogenesis. If parthenogenic animals don't have to bother with finding a mate, then why aren't all animals parthenogenic? Because the offspring of a parthenogenic female are all identical. They are all identical to their mother, since they have no father. That means they all have the same weaknesses and the same strengths. If a hardship comes along—a drought, a famine, a disease, a new parasite—they are all likely to be vulnerable in the same way. If a hardship kills one of them, it is likely to kill them all.

But young that are the product of both a mother and a father are varied. Some are more like Mom, some are more like Dad. They have different strengths and weaknesses. A hardship that kills a few will probably not kill them all. Some are resistant to one thing, some to another. This may not seem very important, but it is. In nature, life is one trial after another. Life is a constant struggle for survival. Variation among offspring greatly increases the chances that at least one of them will survive to adulthood, and that descendants will carry on into the future. This is why the vast majority of animals reproduce by mating.

◄ Insect Disguises ►

Coloring is not the only way animals can be camouflaged. Stick insects are one of the most famous examples of an animal that blends in by shape as well as color. Other animals whose shape helps to hide them are tropical katydids that look like leaves, complete with fake leaf veins, chewed edges, and other flaws that are seen on real leaves. A Madagscan praying mantis looks like a shriveled leaf, including an imitation twisted leaf tip sticking off the top of its head. It even hangs upside down like a dead leaf. Many insects mimic bird droppings, including the rain forest moth *Euprocto*. It is mostly white, with black and brown areas. At rest, it spreads out the silky hairs on its legs to look like splashed areas. Some moths at rest look like very short twigs sticking out from a branch. Treehoppers can look like thorns. Vine snakes look like vines.

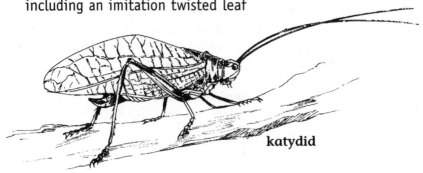

katydid

Where They Live

Stick insects are usually found on trees or shrubs. They are found over most of the United States except for the Northwest. They are most common in the South, but there are a few northern species. Stick insects are most abundant in tropical countries.

What You Can See and Do

If you find a stick insect outside, watch how still the creature is during the day. Notice that the stick insect may sway slightly from side to side by flexing its legs. This helps it to look even more like part of a shrub or tree, because real branches and twigs sway with the wind.

Come back and check on the stick insect as darkness falls. At some point in the evening the stick will begin to move around and feed. It will probably eat the leaves of the plant it is resting on. Wave your hand in front of the stick insect or touch it (not its eyes or head). How does it react? Does it freeze, or drop to the ground? If it is winged, does it rise up and flash its wings at you?

Small wingless stick insects can be kept indoors for a few days for observation. Get a terrarium ready by putting in some leafy twigs from the tree where you found the stick insect. The twigs hold the leaves off the floor of the terrarium—the insect can't eat them if they are lying flat. The twigs also provide something for the insect to climb on. You don't need to add sand, soil, a water dish, or anything else to the terrarium.

You can move the stick insect into the terrarium by letting it step onto your hand. If you must lift it, put a finger gently on either side of its body and pull slowly. Be careful of its legs, which break off easily.

Change the leafy twigs in the terrarium every day. The stick insect can't eat them if they aren't fresh. Spray the sides of the terrarium once a day with water droplets for the insect to drink.

Most stick insects are particular about what type of leaves they eat. If you find a stick insect and you have no idea what tree it came from, you can try offering blackberry or raspberry leaves. Many species in bug zoos will eat blackberry leaves, even if blackberry isn't their favored plant.

If you search for "stick insect" or "walkingstick" on the Internet, you may find Web sites where you can order stick insects through the mail. These sites also provide information on the care and feeding of the stick insects they offer.

Stag Beetles

That's Strange!

The male stag rears up on his back legs and charges toward the other male. His huge branching antlers, over one-third of his total length, gleam in the soft forest light. He crashes headfirst into the other male. Antlers locked, they wrestle to see who will win the female that they both want. Eventually, the weaker of the two males gives up and straggles away. The winner then turns to the female who will become his mate. You may have seen elk or deer act this way on nature shows, but the animals I'm talking about are only a couple of inches long! And the "antlers" are really jaws!

male giant stag beetle

What They Look Like

The male giant stag beetle has jaws that look like two big, forked antlers on the front of its head. The giant stag is the largest species in its family of stag beetles, with a total length of about 2½ inches (6 cm). It is reddish brown and shiny. The male is unmistakable, with his huge forked jaws and wide head. His head is wider than his prothorax, the first segment of the thorax just behind his head. But the female's head is narrower than her prothorax and her jaws are shorter. Her jaws are just barely longer than her head, while the male's jaws are as long as half of his body.

Another fairly common stag beetle is *Pseudolucanus capreolus*, the pinching bug. Its jaws are large, a little longer than its head, but not nearly as long as those of the giant stag. Its body is shorter, too, reaching a length of about 1⅝ inches (4 cm). Like the giant stag, the pinching bug is also reddish brown.

female giant stag beetle

There are still smaller stags, with smaller jaws. They can be distinguished from other beetles by the elbow bend in their antennae and by the leaflike segments or flaps on the ends of the antennae, which cannot be pressed together into a solid ball. (Scarab beetles have similar leaflike flaps on the end of the antennae that can be pressed together to make a clump or club at the end.)

How can you tell that an insect is a beetle and not some other kind of insect? All beetles have two pairs of wings. The outer pair are hard and stiff, and serve as a cover for the inner wings. When the beetle is at rest, the edges of its hard outer wings meet in a neat line down the center of its back.

Why Do They Look Like That?

Why do giant stag beetles have such huge jaws? For the same reasons that male elk, deer, moose, and other hoofed animals have big antlers. In all these species, males compete with other males for a female of the same species.

Males of many different species have a sort of ritual that they go through to decide which one will get a female. Male rattlesnakes raise the top half of their bodies off the ground, wrap around each other, and each tries to pin the other to the ground. This is called a combat ritual. A male lizard of the genus *Anolis* claims a small patch of land for his future bride and then keeps all other males out by displaying a colored throat flap to other males and doing push-ups. A male field cricket will kick, attack, and wrestle any other male field cricket who interferes with his pursuit of a female.

Sometimes when males use a visual signal to scare other males away or to attract females, that signal can become very exaggerated. Some male hoofed animals, such as elk, have antlers that are so big that they can cause problems. Huge antlers are heavy, and in a forest they can get snagged on low branches or vines. But if big antlers are good for defeating rival males, then elk with big antlers will be more successful at attracting mates and fathering babies. They will also pass on that trait to their sons. If every year those with the biggest antlers have the most offspring, then over time the size of the antlers in the population will get bigger and bigger. This is called sexual selection. A trait that exists because of sexual selection does not help an animal survive better, find food better, or escape predators more easily. The trait only helps the animal to attract a mate more successfully.

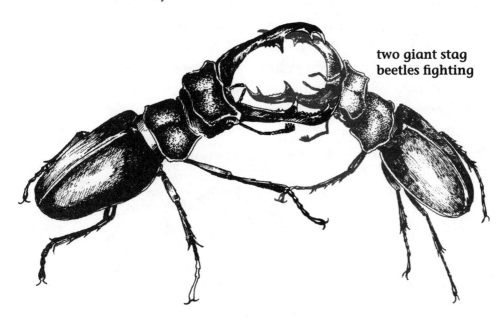

**two giant stag
beetles fighting**

The huge, colorful feathers of a male peacock are a good example of sexual selection. During the breeding season each year, a male peacock grows about 150 beautiful feathers from his lower back. The feathers are 4 to 5 feet (1.2 to 1.5 m) long! When he's relaxing, they form a long train behind him. When he wants to attract and impress a female, he holds his feathers up and spreads them into a fan shape across his back. Then he struts around in front of the female making a loud call. It's a beautiful display. However, the feathers are so big that they probably make the peacock more vulnerable to predators. They certainly don't help him find food. But the male peacock with the biggest and flashiest display is most likely to attract mates (several). So those with the longest and flashiest feathers have the most offspring. As a result, over time their tails have gotten bigger and bigger.

The jaws of the giant stag beetle have evolved in the same way. They are also the result of sexual selection. They are so big that they are probably in his way a lot of the time. But giant stag beetles with big jaws are more likely to get mates, so over time their jaws have gotten bigger and bigger. The jaws of the giant stag don't even pinch very well. Smaller-jawed stags of other species can pinch harder.

Hercules and His Bride

Several other types of beetles that are not in the stag beetle family also have cumbersome headgear as a result of sexual selection. One is the Hercules beetle of Central America. The male is about 6 inches (15 cm) long and has two long horns, one above the other. The top one is about 3 inches (7 cm) long—as long as the rest of its body. Like the giant stag, male Hercules beetles fight each other with their horns, although generally no one is hurt. A male Hercules beetle can pick up a female with his horns and carry her away.

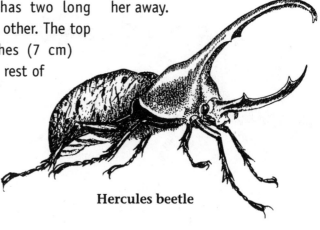

Hercules beetle

Where They Live

The giant stag beetle *(Lucanus elaphus)* is found in the southeastern United States—as far north as Illinois and as far west as Oklahoma. The pinching bug *(Pseudolucanus capreolus)* is in the eastern United States, north to Canada. Other species are found in the West. Stags are most likely to be found in woods, or in open areas near woods. Some species fly toward lights at night.

What You Can See and Do

If you see a stag beetle outdoors, look around to see if there are any others. If there are two males near each other, you may be lucky enough to see them interact. You may see them push against each other with their huge jaws until one of them is flipped over or one gives up and walks away.

Stag beetles are attracted to light, so they sometimes come to porches at night. If you try to catch one, keep your hands away from the jaws.

You can keep a stag beetle indoors for a day or two, in a terrarium with damp sand or damp soil on the bottom. Spray the walls of the terrarium with droplets of water.

Stag beetles eat sap that oozes from trees, so you will not be able to feed one in captivity. Let it go after a day, or two days at most.

Assassin Bugs

That's Strange!

The soldier, dressed all in gray armor, moves with grace and stealth. This warrior moves so slowly and smoothly that its victims seldom see it coming. Ever closer it creeps toward the unsuspecting opponent. By the time the victim realizes that the soldier is upon it, it is too late for escape. The spike of a sword is thrust down and the victim is impaled. Death comes quickly.

If the victim had time to see its murderer, it would see a frightening sight indeed. The soldier is a regal and graceful creature, yes, but not very handsome. Its head is a long gray knob, like a long hot dog or cucumber sticking way out in front of its body. And its sword is not held in its hand, as a sword normally would be. The end of this sword is attached to the underside of the long gray knob of a head. There is no sheath or scabbard for the long blade. Instead the sword is pulled up to rest along the underside of the head when not in use.

Who is this strange soldier? It's a wheelbug, a type of assassin bug. It is perhaps the oddest of the assassin bugs. And all of the assassin bugs are odd.

What They Look Like

All assassin bugs have a long knob-like head with a thin stiff beak attached to the end of it. *Assassin* means "killer or murderer," and all assassin bugs are fierce predators. They have six long walking legs that carry them high off the ground. Assassin bugs are members of the order Hemiptera, which is the group that biologists call true bugs. Other insects are commonly called bugs, but only the hemipterans are true bugs. Hemipterans have distinctive wings that set them apart from other insects. The front, or upper, pair of wings are leathery at the base and membranous at the tip. When folded, these front wings often make a sort of X on the bug's back. The back wings, or underwings, are entirely membranous and are used for flight.

assassin bug

There are many species of assassin bugs, and in many ways they look very different. The wheel bug is one of the largest, up to $1\frac{3}{8}$ inches (35 mm) long as an adult. It is gray and has a curved ridge on its back. The ridge looks like part of a wheel sticking up out of the back. There are points on the ridge, like cogs on a gear or wheel. These points would not be pleasant for a bigger predator to bite down on.

Other types of assassin bugs are often green. As a group they range in size from ½ to 1⅜ inches (12 to 35 mm). The wheel bug and many others are oval shaped, but some have long slim bodies, somewhat like stick insects. Young ones can be tiny. I often see small, very slender green ones that are ½ inch (12 mm) long or so. They look very much like stilt bugs, which are also hemipterans but are plant eaters. The difference is in the shape of the head. Assassin bugs have a long thin head. If you're undecided about whether a land-dwelling hemipteran you've captured is an assassin bug, offer it prey. If the prey is small enough, then the assassin bug will usually eat it.

wheel bug

Why Do They Do That?

Assassin bugs and all the other bugs in the order Hemiptera feed in a very strange way. They all have long stiff beaks that work like sharp drinking straws. They poke the beak into their food and then suck out the food's juices. Many hemipterans feed on plant juices and, for them, dinner is no big deal. The beak pierces the plant stem, sap flows into the beak, and the plant doesn't seem to mind. However, for those hemipterans who dine on other animals, mealtime can be full of drama.

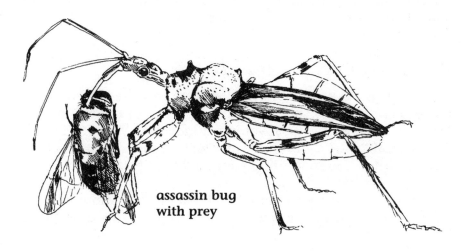

assassin bug with prey

We can assume that any insect who saw an assassin bug coming would probably try to get out of the way. The stealthy assassin bugs have made allowances for this. They creep up on their prey very much as a tiger does. They glide very smoothly, very slowly, until the last moment. Then all of a sudden, the long front legs dart out like a striking rattlesnake. Wham! The prey is totally surprised. The prey doesn't run or yelp because it never has a chance to.

The two front legs of assassin bugs are often different from the four back legs. The front legs are often thicker and many have spines to keep the prey from slipping away. Once the front legs are folded around the prey, the victim seldom moves again. Then comes death. The free end of the long beak swings down from its resting place against the assassin bug's "chest." The beak easily pierces the prey, even the hard exoskeleton of a beetle.

The assassin bug usually remains completely motionless for the next hour or two, as it sucks its dinner out in the same way that you might suck a soft drink from a glass using a straw.

Where They Live

Wheel bugs are found throughout the states east of the Rocky Mountains and in southeastern Canada. Other kinds of assassin bugs can be found both east and west of the Rockies, throughout the United States and southern Canada.

If you are looking for an assassin bug, you may have to wait for one to show up. They are usually green, gray, or brown, and they spend a lot of time sitting still. This makes them very hard to see in nature. Most of those I've found have been either in my house, on my deck, or on a sidewalk, where their camouflage did not work for them.

Outside, they are usually on plant leaves or on the ground. You might find one by spreading a sheet or umbrella on the ground under a bush or small leafy tree and shaking the bush or tree vigorously. Something will fall out—it could be an assassin bug. You could also find one by sweeping a net through dense leaves. Sometimes they are near gardens, feeding on the insects that are eating the vegetables.

What You Can See and Do

Assassin bugs will not run or fly if you try to get close to them. They have no need to run—they can use that sharp beak for protection as well as feeding. *Don't pick one up.* If you do, it may poke you. The beak injects substances into prey that dissolve prey on the inside. The assassin bug will inject the same substances into your finger if it pokes you, and you will feel a painful sting.

I have never seen one jump and I don't recall seeing one fly unless someone was poking at it. But even if it did fly, an assassin bug would never fly right at you for the purpose of stinging or biting you, as a wasp might. Instead, it would fly away.

If you see one outside, you can offer it water. In nature, they often drink from dew droplets. You can put a big drop of water in front of your assassin bug, and if it is thirsty, it will put one end of its beak in the water and drink.

You can also offer it prey. A very small assassin bug, ½ inch (12 mm) long or so, will be happy to eat aphids. Aphids are pear-shaped soft-bodied insects of various colors (red and green are common) that live on the underside of leaves on garden plants, like lettuce and tomato. You can also offer a small caterpillar to a small assassin bug.

If you find a larger assassin bug, like a wheel bug, you can offer it larger prey, such as a big caterpillar or a mealworm beetle. I've watched wheel bugs eat mealworm beetles many times. The excitement begins as soon as the wheel bug sees the beetle move. The predator turns toward the beetle and slowly walks over to it. Without the slightest hesitation or fear, the wheel bug enfolds the prey in its two front legs so the beetle can't move. Then, ruthlessly, the wheel bug stabs the prey with its beak—just once. The beetle, unable to move, doesn't struggle. The wheel bug may hold the beetle like that for more than 2 hours while it drains the creature's insides. When the meal is over and the beetle is dropped, its body is like a Rice Krispie. It's completely hollow, although it looks the same from the outside. If you hold it up to the light, you can see its insides have been completely dissolved and removed.

Dung Beetles

That's Strange!

Before you were born, your parents prepared a cozy little bed for you where you would be safe and warm and dry. After you arrived, they watched over you and protected you. They fed you milk for the first several months of your life. The milk was warm and sweet and provided all the nutrients you needed.

The beginning of life is very similar for a young dung beetle. Its mom and dad prepare a safe and dry place for it to rest, too. They watch over their baby and keep it safe from harm. And they provide for the little beetle larva just the perfect food. It's warm and soft and goes right down the hatch. This special food will provide all the nutrients that their beetle baby needs—exactly the right thing for their precious larva. The special food they prepare is a nice, fresh ball of dung. That's *dung*, as in cowpat or cow patty—cow poop. The fresher, the better.

What They Look Like

A dung beetle is a beetle that eats dung and feeds dung to its larvae. This includes many different species of beetles. All of them are related. All are in the family Scarabaeidae, the scarab beetles, or scarabs. They are all oval shaped, usually thick and heavy bodied. The antennae of scarabs are very distinctive. At the tip of each antennae is a knob rather like the end of a golf club. The knob can at times separate into several flat plates, as though it were cut into slices. These flat plates can pick up scents from the air. Many dung beetles are dull black or green. Black ones may have a greenish or copper sheen. Some have horns or spikes or other odd shapes on the head or pronotum (just behind the head). These shapes aid in digging or moving soil, or in cutting dung. Many dung beetles have a spur on each of the two back legs that help in moving balls of dung. The scarab beetle family includes many that are not dung beetles, such as green june beetles, rhinoceros beetles, Hercules beetles, and Japanese beetles.

sacred scarab

Why Do They Do That?

Eating dung is strange, but it's not the very strangest thing about dung beetles. What's more remarkable is the way they care for their young.

Most insects pay no attention to their young. Mom and Dad mate, Mom lays the eggs, and neither parent ever sees the young or each other again. This is typical for most animals on the planet, especially invertebrates (animals without backbones) such as worms, jellyfish, clams, sea urchins, crabs, and so on. Fish, amphibians, and reptiles are **vertebrates** (have backbones) like ourselves, but they usually abandon their young, too. Birds and mammals are the only two major groups of living things that uniformly provide some care for their young. Every bird and every mammal species protects their eggs or young in some way. And one or both parents usually feed the young. Some birds such as ducks and chickens and turkeys may not feed their young but one parent at least protects the young while they feed themselves.

This kind of attention in animals is called **parental care.** Animals with parental care usually have very few offspring. The fewer babies you have, the more attention you can give each one. With parental care, each baby has a high chance of survival. Animals that have no parental care, like flies, clams, and fish, for example, may lay hundreds or thousands of eggs. Most of them will die but a few will survive. We know that both strategies can work, because we see the results all around us.

Parental care in the Egyptian sacred scarab (*Scarabaeus sacer*) was studied at great length by a famous French entomologist named Jean-Henri Fabre. An entomologist is a scientist who studies insects. Fabre learned about these dung beetles by lying on his stomach in a pasture for hours and watching the beetles at work. He saw that the mother beetle makes a ball of cow dung and rolls it away from the cowpat to a more private spot. The ball may be as large as a baseball. She pushes it with her back legs while walking backward. The spurs on her back legs help her hold on to it. When she finds a spot of soft soil, she digs a hole, then pushes the dung ball into the hole. If she digs a hole very close to the cowpat, she may just move small pieces of dung into the hole, then form the ball inside the burrow.

She climbs into the hole or burrow, too. Then she works on the ball, changing it to more of a pear shape. At the very top of the narrow part is a little crater, where she deposits a single egg. The mother beetle then covers the egg with loose fibers of dung. She covers the opening with soil, and then leaves to do it all again. She will make five or six such burrows, each with a single egg.

After a week to 10 days, the egg hatches. The small grub (wormlike beetle larva) begins right away to eat the dung inside the ball. The outer part of the ball has formed a sort of crust, which is not eaten. The grub

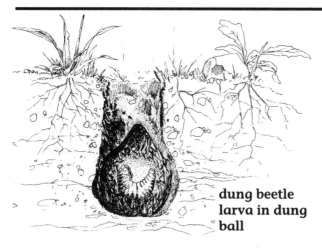

**dung beetle
larva in dung
ball**

grows big and fat, then becomes a pupa. In August or early fall, rains come which soften the walls of the dung ball underground. Then an adult beetle crawls out of the pupal skin and out of the dung ball, up to the surface of the ground.

Fabre discovered that adult sacred scarabs make dung balls to feed themselves as well as their young. Each newly emerged beetle rolls its dung ball away from the cowpat and digs a hole. It climbs into the hole with the dung ball, shuts itself in the burrow, and stays there until the ball has been eaten. Then the beetle comes out of the hole and goes to make another ball.

Another type of dung beetle is *Copris hispanus,* which lives in countries in Europe near the Mediterranean Sea. Its lifestyle is similar to that of the sacred scarab except that the young get more care from their parents. The male carries away the soil as the female digs a tunnel. Then they both move sheep dung into the tunnel. The female presses it all together into a patty, and the male leaves. The female makes three or four balls the size of golf balls, from the dung in the patty. She lays a single egg in a small crater on each one. The mother beetle then guards the eggs for 3 months, with nothing to eat for herself. She keeps the balls fresh by cleaning off sprouting seeds or fungi, and repairing cracks. When the young have matured into beetles, she escorts them outside into the wide world.

In the United States, there are dung beetles called tumblebugs, in the genus *Canthon.* The male tumblebug often helps the female form the ball of dung and roll it away from the cowpat. He pulls on it while she pushes with her back legs. Both walk backward. The female may be helped instead by some other adult, who will try to steal the ball of dung while she is busy digging the hole. The female tumblebug lays a single egg on a ball of dung, then covers it with soil and leaves it alone.

**male and female
tumblebugs
moving dung
ball**

Tumblebugs and other dung beetles perform a very important service. For one thing, they clean up the big piles of dung that would otherwise accumulate in pastures. After the dung is broken into small pieces, wind can blow it away. But dung beetles also enrich the soil by burying pieces of dung, which is the greatest kind of fertilizer. As one researcher in Canada said, "If it weren't for dung beetles, we'd be up to our eyeballs in you-know-what."

◄ Let's Grab Some Dinner! ►

Not all dung beetles eat cow dung. Many specialize in the dung of much more exotic mammals. Some hang on to the hind-end fur of their particular animal—a sloth or kangaroo—and wait until the animal needs to relieve itself. Then they jump aboard the droppings. Some ride all the way down from the treetops on the just-ejected dung of howler monkeys.

Where They Live

Dung beetles live throughout most of the United States and Canada, and in most areas of the world where there are pasturelands. You can find the beetles in areas where large animals graze. Look under and around fresh piles of dung.

What You Can See and Do

You may be able to find a burrow where an adult is feeding or where an egg is hidden in a ball of dung. Look for a small mound of loose soil, wide enough to cover a fist-size hole. If you lift the soil cover very gently, you may be able to observe a beetle. Maybe it will be working on the ball, preparing it for egg laying. Or maybe it will just be eating its own dinner. It may allow you to watch if you sit very still. You may observe a ball with no beetle in sight. Maybe an egg, larva, or pupa is inside.

If you see a beetle moving a ball, you can do a little experiment to see if it can solve problems. Use a rock or a couple of pencils stuck in the ground to block the path of the ball. Can the beetle change direction to go around the obstacle? What if you put two or three rocks in the way? Can the beetle turn around and go in the other direction?

If you touch the beetle or the dung, be sure to wash your hands carefully afterward.

Robber Flies

That's Strange!

You're a honeybee, out with your sister bees collecting pollen. *What a bright sunny day*, you think to yourself. What a happy day to be alive, to be flying around in the warm air, to be a bee. Oh! Here comes a new bee! You wonder, *Does she belong to our hive?* Hmm. Something is not quite right about this new bee. What is it? Oh well. She's out of sight for the moment. You decide to visit one more bright orange flower before heading back to the hive to unload your pollen. Your sister is just ahead of you. Then, out of nowhere, *wham!* What's going on?! Your sister has been slammed in midair by the strange-looking new bee! That's not nice! What kind of bee behavior is this? Wait! This must be an ugly joke. The new bee has wrapped its long legs around your sister and is now jabbing her with a dangerous-looking sharp mouthpart! That jabbing mouthpart doesn't look very bee-like. Those long bristly legs don't either. As you watch in horror, your sister slowly stops struggling and the odd new "bee" carries her away.

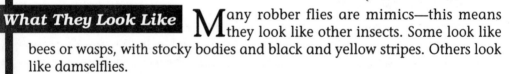

robber fly

What They Look Like

Many robber flies are mimics—this means they look like other insects. Some look like bees or wasps, with stocky bodies and black and yellow stripes. Others look like damselflies.

There are over 4,000 species of robber flies worldwide, and many of them are not mimics. Many of those that don't look like bees have long slender abdomens. Most have a large thorax and long legs. The large thorax and long abdomen together can make them look like little helicopters. Their length varies, from ¼ to 1⅛ inches (5 to 30 mm). Robber flies can turn their heads, a trait that they share with only a few other predatory insects, such as praying mantises and dragonflies.

The entire body of a robber fly is often hairy or bristly. (The hairs are not true hairs. Only mammals have true hairs. Rather they are bristles that look like hairs.) On the head of most robber flies is a thick beard of bristles. All robber flies have a hollowed-out area between the eyes, which other flies do not have. Like all flies (all members of the order Diptera), robber flies have only two wings.

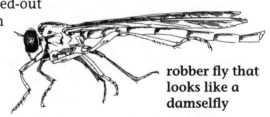

robber fly that looks like a damselfly

Why Do They Do That?

You can probably guess why some robber flies look like bees or wasps. Bees and wasps are easy to recognize and have a painful sting, so not many predators eat them. The resemblance probably helps robber flies avoid being eaten. But robber flies have an additional benefit from looking like bees. You know the old story about the wolf that dresses up like a sheep in order to get closer to the sheep, so that it can eat them. The robber fly that mimics a bee is doing the same thing. Looking like a bee may help it get closer to its prey. Some robber flies even lay their eggs in bees' nests. When the robber fly larvae hatch, they eat the bee larvae. Tricky!

Robber flies' method of catching prey is very similar to that of dragonflies. Like dragonflies, robber flies catch other insects that are flying in midair. To do this, they either fly or leap from a perch that is well off the ground. It's not easy to fly in a straight line to a moving target. You have to be able to predict where the target is going to be a few seconds *after* you take off. It's like throwing a ball to someone who is running. The robber fly must solve this problem when leaping or flying to catch prey. Yet its brain is probably no bigger than the head of a pin. That's a big calculation for so small a creature.

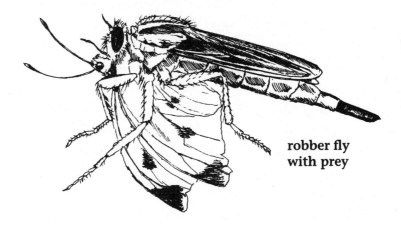

**robber fly
with prey**

After the robber fly intercepts the prey, it must subdue the struggling insect. While still in flight, the robber fly wraps its legs around the prey—a bee, wasp, fly, beetle, or even a grasshopper or dragonfly. Then, still flying, it pokes the prey with its short, sharp proboscis, a mouthpart like a drinking straw. A chemical comes out of the proboscis that paralyzes the prey, so that the victim stops moving in 30 to 50 seconds. The robber fly returns to its perch to eat in comfort. With the poking mouthpart still in place, it now injects digesting chemicals into the prey that turn the creature's insides into liquid. The robber fly slurps them up with its sharp, hollow proboscis.

How to Find That Special Someone

Robber flies have yet another oddity—the courtship behavior between the sexes. **Courtship** is the behavior between mature male and female animals that communicates their species identity and their interest in mating. A male bird that is interested in mating with a female bird might signal to her by bobbing his head in a particular way, flapping his wings, or moving his feet in a sort of dance. Or he might sing a specific song to her or feed her in a particular way. If the female is interested in him, she might respond with some other specific movement that would encourage the male.

We seldom see courtship in insects, although there are a few examples that are very well known. The lights of fireflies are male fireflies signaling to females. Courtship may have developed in robber flies because the female has a tendency to grab nearby insects and eat them. The male has reason to be concerned! By signaling the female before he gets too close, he lets her know that he is a male with romance on his mind. The exact motions and sounds of the signal vary from species to species. Some species of robber flies signal by making buzzing sounds at a certain pitch. Visual signals might consist of the male kicking his legs, or flashing his wings to display a handsome black-and-white pattern. Some males even give the female a present, which distracts her. The present may be an insect wrapped in silk from his leg glands. A tricky male may even give his sweetheart a silk package that turns out to have nothing inside.

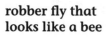

robber fly that looks like a bee

Where They Live

Robber flies are found more or less throughout the United States and Canada, especially east of the Rocky Mountains.

Robber flies tend to be found in and near pastures, meadows, and fields. Some species are more likely to be found near bees' nests and/or flowers that bees visit. If you are looking for one, look on the tips of leaves and the tips of twigs on the edges of fields or flower gardens. Look also on grass stems and on bare patches of ground. These are the kinds of places a robber fly might use for a perch. From its perch, it can keep an eye out for passing insects. You will see it crouching on the leaf or twig, its bearded face looking out and its front legs drawn up. It is ready to spring, and grab some little bug that's not paying attention.

What You Can See and Do

Robber flies often fly or spring from their perches just to check out insects or objects that go whizzing by them. They can't see very well, and have to get close to something to tell what it is. So most of their flights are just to investigate. You may be able to get a robber fly to leave its perch and fly out by playing a trick on it. You can toss a small piece of twig or a small pebble over or in front of a robber fly. The fly may leave its perch to see what it is. If not, you may get to see the robber fly move its head to follow the direction of the object you threw. The head turning is important in allowing the robber fly to track the path of potential prey, and to fly off in the right direction to intercept it.

Since robber flies have a sharp, jabbing mouthpart, don't touch them or pick them up.

Giant Water Bugs

That's Strange!

You're walking along the edge of a pond. Every once in a while you disturb a small frog perched on the bank or in the shallow water. Afraid of you, it leaps into deeper water for safety. As you walk, you enjoy the plopping sounds as each one in turn smacks the surface. But then you come to one that doesn't leap. It just sits there in the shallows. You stoop to look at it closely and notice that its skin is beginning to collapse like a deflating kick ball. Puzzled, you stare at it. The frog continues to shrink as you watch. Soon its empty skin floats in folds on top of the water.

Annie Dillard in her book *Pilgrim at Tinker Creek* describes this encounter with a deflating frog. What was going on? When she looked behind the frog, she saw the brown oval body of a giant water bug in the water. The water bug was sucking out the frog's insides.

What They Look Like

Giant water bugs look rather like roaches—flat, brown, and oval. But they live in water instead of on land. Giant water bugs look a bit like beetles as well. This can be confusing, because some beetles live in water and they may also be brown and oval. But giant water bugs are much flatter than beetles. They also differ from beetles in the way the wings come together at rest. A beetle's wings meet in a straight line down the middle of the back. Giant water bugs' wings are different. Because giant water bugs are hemipterans, their folded wings create an X pattern on the back, or at least part of one.

giant
water bug

Hemipterans are the order of true bugs. Other insects may be called bugs casually, but to insect scientists, only those in the order Hemiptera are true bugs. The order also includes water striders, assassin bugs, milkweed bugs, back swimmers, and water boatmen. All of these have an X pattern on the back because the wings cross at rest and the back half of each wing is clear. Hemiptera means "half wings" (*hemi* means "half" and *ptera* means "wings"). The X pattern is more obvious on some than on others.

There are many aquatic hemipterans, and they're nearly all predators. How can you tell a giant water bug from the others? Giant water bugs are the biggest of this group. Adults range from 3/4 to 2 1/4 inches (25 to 55 mm) long, making the giant water bugs among the largest insects in North America. The very largest ones—

those that are longer than an inch—belong to the genus *Leucocerus*. Those of the genus *Belostoma* are somewhat smaller as adults, an inch or a little less in length.

The giant water bug has some interesting adaptations for swimming and catching prey. It has a keel on the underside of its body. This helps it move through the water in a straight line. The keel of a boat has the same function. The four back legs of the giant water bug are also built for swimming. They are flattened to work as oars, and have a fringe of hairs on the side to make the oars even wider. Some have more hairs than others (see drawing below).

The front legs of the giant water bug are sharp and hooked for grabbing prey. The mouth is a powerful sucking beak that stays tucked up against the "chest" when not in use. Hemipterans don't have a larval life stage. Instead, the eggs hatch into nymphs, which are small wingless copies of the adults.

Why Do They Do That?

A giant water bug doesn't look frightful, but if I were one or two inches tall, I would do anything to avoid the giant water bug. This bug would be my worst nightmare.

It is a very powerful predator. Not many predatory insects can eat animals that have backbones, like amphibians and fish. But giant water bugs routinely eat small animals with backbones (vertebrates), as well as anything else they can grab. Like all hemipterans, they have a beak that sucks up food. First they grab the prey. Then they jab the stiff beak into the victim. The bugs inject a substance into their prey that turns the victim's insides into a liquid. Then they suck out the juice. If a giant water bug poked your hand with its beak, it would feel like a sharp sting and your hand would hurt or feel numb for several hours. *So do not pick up these guys.*

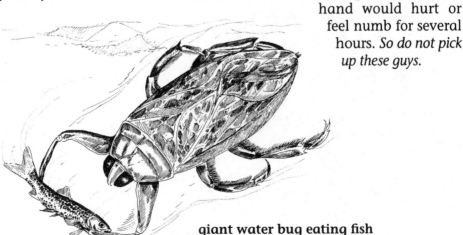

giant water bug eating fish

Giant water bugs are strange in another way, too. One of the first ones I saw had what looked like a bunch of long white cocoons standing on end across its back. The white things were eggs, giant water bug eggs. You might think that the bug who carried them was their mother. But it was their father—an unusual situation in the insect world. Male giant water bugs with eggs on their backs are at much greater risk of being eaten by fish or birds because the eggs slow them down. Also, the eggs make the adult bug more visible to predators. So why does the male carry the eggs around? For one thing, he keeps the eggs safer from predation than they would be on a rock or a log—unless, of course, Dad himself gets eaten. But he is less likely to be eaten than the eggs alone because he can move around to avoid predators.

How do the eggs get on Dad's back in the first place? Immediately after the male and female mate, the male guides the female into position on his back and she begins laying eggs. She starts at the rear and moves forward, until his back is more or less covered. From then on, Dad's in charge.

Not all male giant water bugs have this fatherly duty. Those in the genus *Belostoma* do, but females of *Leucocerus* lay their eggs on aquatic vegetation. In this case, the fathers have no parental duties.

Giant water bugs breathe in a strange way, too. They breathe air even though they live in water. This is true for marine mammals such as whales and porpoises, too. Whales and porpoises have a blowhole on the back that they breathe through when they come to the surface. Giant water bugs don't have blowholes, but their solution to the problem is not so different. In order to breathe, they come to the surface, rear first, and take in air through two tubes that stick up above the surface.

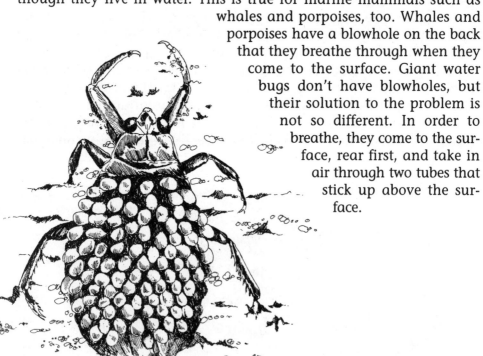

**giant water bug with
eggs on back**

Dad Rocks the Little Ones

A scientist named R.L. Smith did some experiments with male giant water bugs that were carrying eggs on their backs. He discovered that there's more to it than just protection from predation. He found that eggs he removed from the dad and placed in the bottom of a dish of pond water didn't hatch. Then he noticed that Dad usually keeps the eggs right at or just below the surface. Do they need a little bit of air? He removed eggs from the male again and kept them right at the water's surface. This time some of them hatched, but still most did not.

Smith noticed that the males of some giant water bug species often stroke the eggs with their back legs. Males of other species rock the eggs back and forth by pivoting on their middle legs—head up and tail down, then head down and tail up, and so on. This fatherly attention seems to be essential to the eggs' development, because they won't develop without it.

Where They Live

Giant water bugs are found throughout the United States and Canada. They are found in shallow ponds, or in the shallows of larger ponds. Usually they stay hidden in aquatic vegetation.

What You Can See and Do

The first time I saw a giant water bug was when I ordered three from a biological supply company (see Resources). I received a box in the mail. Inside the box were pieces of foam packing material around a plastic bag of air and water. I pulled the bag out, and in it were three giant water bugs, alive and well, floating on the surface of the water. There are many different species of giant water bugs, and this was a small species of *Belostoma,* only about $3/4$ inch (18 mm) long. I was a bit disappointed that they were so small. I wanted to use them for a science class and I wanted something huge and dramatic.

I settled them into the aquarium I'd prepared, then put one into a smaller plastic aquarium and put it on the kitchen table in front of me, so I could keep an eye on it while I did some paperwork. Just to see what would happen, I put in a couple of live fish I had from a different aquarium—a guppy and a goldfish. The guppy was a little less than 1 inch (2 cm) long, the goldfish about 2 inches (5 cm) long. Within a minute, the bug swam over to the guppy and grabbed it with the two hooked front legs. The guppy was totally pinned—it could barely wiggle. The giant water bug swung out its long curved beak and poked it into the fish. That was it for the fish. The bug clutched it for an hour or so, floating on the surface and sucking out the fish's body fluids. When the bug finally released it, the fish looked the same, maybe a little thinner. But it was most definitely dead—and empty. I no longer felt disappointed in the size of the bugs. I was sure they would provide plenty of action and drama for my science class. And they did.

You can order giant water bugs through the mail, too, if your parents agree. (See Resources on page 103.) Put them in a clean aquarium, with the water they arrived in. If you need to add water, use pond water or bottled spring water of the same temperature. Keep a lid on the aquarium because these insects can fly. Remember, don't touch or handle them—they bite! They will eat aquatic insects, small fish, or tadpoles.

You may be able to find one in nature. Follow the same procedure described for catching back swimmers (page 35).

Part III

Strange Insects You Probably Won't Find

Leaf-Cutter Ants

That's Strange!

You are walking along the sidewalk at La Selva Biological Station in the lowland rain forest of Costa Rica. You're enjoying watching the family of peccaries wandering among the few small research buildings in a clearing next to the forest. The peccaries look a lot like their pig relatives, but smaller and hairy and gray. You see up ahead an agouti crouched in the grass. The house-cat-size rodent reminds you of a big squirrel without the bushy tail. Over to the left, a raccoonlike coatimundi walks out of the forest, sees you, and slips back in.

Watching the coati, you almost miss the narrow river of green flowing across the sidewalk. You look closer at the green stuff and see that it's pieces of green leaves, thousands of them, all fingernail-size. They appear to be floating in air, just off the ground, lined up to make a narrow ribbon that continues in both directions as far as you can see. To the right, the green ribbon crosses the lawn for more than 100 feet before entering the jungle. There is a brown path directly under the ribbon of green, all the way to the forest edge. You follow the narrow path, walking beside it, until it disappears into the jungle. Then you turn around and follow it back to the sidewalk and beyond, until the channel of green leaves enters a hole in the ground.

leaf-cutter ant

As you stoop to look closer, you see that the leaf fragments are not floating after all. They're being carried. The river is really a thick line of ants. Each ant holds a piece of leaf over its head like a parasol or a flag. Why? Why are they taking these leaf fragments into a hole in the ground? To eat them? No. They are feeding the leaf pieces to their captive fungus.

What They Look Like There are some 35 species of leaf-cutter ants, and they don't all look identical. Several are in the genus *Atta*. In general they are reddish brown ants, with longer legs and bigger heads than most. Their size varies, depending on their job within the colony. Some are less than 1/10 inch (2.5 mm); others are 1/2 inch (12 mm) in length.

In the tropics, you will recognize leaf-cutter ants by the bits of green leaves they carry over their heads. It's impossible to confuse them with any other ants because no other ants do this.

Why Do They Do That?

Ants are social insects, which is strange enough in itself. Social insects live in large groups or colonies. A colony is a group of creatures of the same species that live together. Colonies of leaf-cutter ants can contain over 1 million individuals.

The strangest thing about a colony of social insects is their division of labor. Division of labor is a common thing in human societies. It means that adults have different jobs. Some grow food. Some make clothes. Some build houses. And the benefits of all these jobs are shared. But division of labor is rare among animals.

Social insects, especially ants, are well known for their complex societies. In any given ant colony, some adults are **soldiers**, which defend the colony. These are the largest (other than the queen, who may be as big as a finger). Some adults are workers, which hunt outside the nest for food and bring it back. Other workers stay inside the nest and care for the immature ants—the eggs, larvae, and pupae. Only one adult in the colony, the queen, has the job of producing eggs. Very large colonies may have more than one queen. But even so, there are hundreds or thousands of workers for every queen.

In a leaf-cutter ant colony, there is even more division of labor than in most ant societies. That's because leaf-cutters have so many jobs to do. The jobs are all about the leaves—what are they doing with those leaves?

In the 1800s, the few scientists who visited the tropics made guesses about the purpose of all those leaves being carried into the ants' colonies. Some of them guessed that the ants were eating the leaves. That's a good guess. Insects and other animals can spend a lot of time gathering food.

Other scientists guessed that the leaf-cutter ants were using the leaves to plug the holes that each nest has in its roof. The roof of a nest has uneven piles of soil on top, soil that has been carried out to make the chambers and tunnels underground. There are hundreds of holes in the top and around the edges for ventilation. It makes sense to think that the

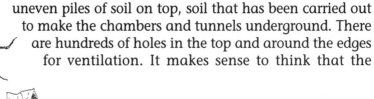

line of leaf-cutter ants

leaves might be used to plug these holes and keep the heavy rain in a rain forest from flowing into the holes. But that's not what the leaves are for.

A leaf fragment begins its journey when a leaf-cutter worker bites it off of a whole leaf on a forest shrub or other plant. She carries it back to the colony in her jaws with the parade of coworkers, each carrying a piece of leaf. After she enters the nest, she follows a tunnel down into a chamber, one of many chambers. There she drops the leaf piece, communicating with a different sort of worker who has come into the chamber. Ants communicate with chemicals. Every ant can secrete up to about 10 different chemicals, from either the abdomen or the head, especially near the mouth. Other ants taste the message with their mouths, or smell it with the odor detectors on their antennae. That's how ants say to each other, "Take over this piece of leaf!" or "We are being invaded" or "Follow this chemical trail to a new bush."

This other worker in the underground chamber gets the message that a new leaf piece has arrived. She is a little smaller than the one who brought the leaf piece. Her job is to clean the leaf piece and chop it with her jaws into small pieces the size of a sesame seed. Then even smaller workers come to chew the pieces into damp pellets.

If you were able to see into this chamber, and many other chambers in the same nest, you would see a spongy, grayish white mass—a big ugly gob of something that smells like mold. The lives of the leaf-cutter ants revolve around this soft, shapeless stuff. They nurture it, protect it, and provide everything it needs. It is a fungus. When the small workers finish making the pellets, they give the pellets to the fungus. They tuck the wads of chewed-up leaves into the mass of fungus. Within a day, the fungus grows over the new leaf pellets and begins digesting them. So this is what the leaves are for—they are food for the fungus. But why would ants take care of a fungus? Because the ants eat this fungus. In fact, it's all they eat.

The fungus engulfs the pellets so rapidly that the ants are constantly digging new chambers for it. As the fungus fills one chamber, another group of workers—the transplanters—cut pieces of it and carry the pieces to new chambers to grow more. The ants are like farmers. They have to prepare a spot for the fungus, provide the nutrients it needs, fertilize it (with their poop), harvest it, and start new growths.

The fungus that leaf cutters grow is not just any fungus, but a very special one that these ants have been culturing for millions of years—about 25 million years. This particular fungus, spread over thousands or millions of leaf-cutter ant colonies, has become so dependent on the ants that it has lost the ability to reproduce with spores, as fungi normally do. It reproduces only when ants break off pieces of it and carry the pieces to new chambers, or when a new queen leaves the colony and takes a piece of the fungus with her, carried in a pocket in her mouth. Why should the fungus waste energy making spores, when the ants will do the job instead? (A spore is a tiny dustlike speck that can grow into a new fungus.)

The fungus and the ants have become so suited to each other that the fungus produces special bite-size body parts just for the ants to eat. They look like tiny cabbages and are called gongylidia. The workers and soldiers eat them, and the workers feed them to the queen and the ant larvae. Actually, they place the larvae in the chamber with the fungus so that they can reach it on their own. The larvae eat almost constantly.

This special fungus must be protected from competing fungi, whose spores might be carried in from the forest by accident. Very small workers, the size of tomato seeds, crawl into narrow spaces in the fungus garden and weed out other types of fungus, just as human farmers weed out unwanted plants.

◀ Riding Shotgun ▶

The most interesting of all the jobs in a leaf-cutter colony is that of another tiny worker. This one goes out with the larger workers or scouts—those that find plants, cut leaf pieces, and carry them back to the nest. To find this tiny companion, you would have to look carefully at the leaf pieces as they are carried in the parade of green from plant to nest. This tiny ant rides on top of the leaf pieces! Is she just lazy? No. She's riding shotgun, like the guys with guns who used to ride on top of a stagecoach in the old West. They were looking for bandits, protecting the stagecoach. This little ant is doing the same thing, protecting her big worker sister from flies that try to attack her. Certain flies in the family Tachinidae try to inject their eggs into the worker ants. If they succeed, the fly egg will hatch inside the body of the worker ant and eat it and kill it. The worker carrying the leaf is unable to defend herself, so the tiny ant chases the flies away from her. She runs around the leaf parasol and snaps at the flies, driving many of them away.

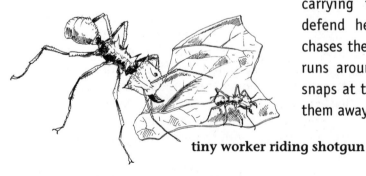

tiny worker riding shotgun

Where They Live

Leaf-cutter ants live in the tropics and subtropics of North, South, and Central America. This includes the southern margins of the United States. Their colonies are in forests or in clearings next to forests, in soil that is well drained, not swampy.

A well-established, mature colony can be as big as 50 feet (15m) across on the surface. But a new colony may appear on the surface as just a hole in the ground, as big around as a human arm.

What You Can See and Do

If you live in Louisiana or Texas, or if you go south of the United States, you may discover a parade of ants carrying green parasols. First of all notice that ants are going both directions on the trail. Those heading toward the nest are carrying leaf pieces. Those heading away from the nest are not. They're following the chemical trail laid down by their sisters. It's leading them to a plant with usable leaves.

Look very carefully and you may be able to see a little ant riding shotgun on some of the leaf pieces. Maybe you'll even see a tachinid fly bothering the workers.

The stream of ants may go on for hundreds of feet. Follow the ants without leaves, to see if you can find the plant they are heading toward. Watch them climb the plant. Notice how they use their jaws to snip pieces of its leaves and carry them to the ground.

Then follow the trail in the other direction to see if you can find the nest. A fairly new nest may just be a hole in the ground, but an older one will have mounds of soil on top, like drip castles.

See if you can get a few soldier ants to come out of the hole. Rake the ground gently with a twig. Or use a leaf to push some of the workers gently off the trail. This will probably cause them to release alarm chemicals, which will bring the soldiers running. If these two methods fail, try putting some other kinds of ants onto the trail right next to the hole. See what happens. Do the leaf-cutters welcome their visitors? *If soldiers appear, don't let them get on your shoes and don't touch them. They bite.*

Botflies

That's Strange!

The researcher was relieved to be out of the rain forest. He'd gotten lost while trying to track down some rare orchids. The early rains had flooded a couple of creeks and cut off the paths he knew. Then he'd lost his compass. But after 6 days of stumbling around in the Bolivian forest and sleeping on the ground, he'd made it to a riverbank and a boat had picked him up. Now he was trying to get settled in and get back to work at the research station. Everything was fine. Except for that lump on his neck. Just some kind of bug bite, he thought. Or maybe an allergic reaction to some poisonous plant. But when he showed the growing lump to his assistant, Juan knew at once what it was. He'd grown up here and he knew all the perils of the forest. "It is *el tórsalo*," said Juan. "*El tórsalo* lives in the lump. And he won't come out until he's ready. Unless you have a beefsteak . . . then maybe he will."

What They Look Like The human botfly is known in some parts of South America as the *tórsalo*. Its scientific name is *Dermatobia hominis* (*Derm* means "skin" and *hominis* means "human"). It is an apt name, because the larva spends its life in human skin. It's not dangerous, but it can be a nuisance.

The larva starts out quite small. It comes from an egg no bigger than the period at the end of this sentence. When it is fully grown, it has the shape of a fat, white worm, about 1 inch (2.5 cm) in length. Like most fly larvae, it is legless.

adult botfly

The parent botfly looks like a common blowfly, bluebottle fly, or housefly. It has the same general shape as these, but bigger—½ to ¾ inch (13 to 19 cm) in length. All flies are in the order Diptera. *Di* means "two" and *ptera* means "wings," an appropriate name because flies have only two wings. Most other insects have four. Adult flies usually have big compound eyes, which is part of what gives them their characteristic look, especially in cartoon drawings and animated movies. Most of them have tubular mouthparts for sucking, not chewing. You can easily watch a housefly suck up a droplet of sugar water on your kitchen table. Some flies have mouthparts for piercing. It may surprise you to learn that mosquitoes are flies, although they don't have the typical fly body shape. Mosquitoes, as we all know from experience, have mouthparts that are for piercing.

Why Do They Do That?

Botflies are **parasites.** A parasite is an animal that feeds and lives on another animal, without killing it—or at least not right away. The animal that is lived and fed on is called the **host.** You know about some common parasites, although you may not associate them with that word. If you have a dog or a cat, you know about fleas and ticks. These are external parasites, meaning that they live and feed on the surface of the body. They suck blood through the skin of their host.

You might know that dogs and cats and many other animals can have intestinal parasites, too, such as tapeworms, hookworms, or round-worms. These parasites live inside the intestines of their hosts and feed on the food that the host has eaten, as the food passes through the intestines. Intestinal parasites don't eat the host's flesh or body, they just steal the host's food. They can make your dog or cat sick by robbing it of nourish-ment, which is why we have pills to kill intestinal parasites.

There are thousands of different kinds of parasites, both plants and animals. Most types of animals have parasites of some kind. Even humans have them. Humans can have external parasites, such as head lice. You may have known someone at school who had head lice, or you may have yourself. They are fairly common. Humans can also get intes-tinal parasites. Fortunately, we have medicines to get rid of them. In parts of the world with little medical care and poor sanitation, humans can and do have quite a variety of parasites.

How do botflies get their eggs under a person's skin? If the parent fly is ½ to ¾ inch (12 to 18 mm) long, and if it makes a buzzing sound like most flies do, why doesn't it get smacked or chased away? The adult botfly avoids this problem by getting a smaller creature to help. It grabs a mosquito or some other small bloodsucking insect, sometimes in midair. The botfly then glues its eggs to the underside of the mosquito. The loaded mosquito sooner or later lands on a warm body, looking for a blood meal. It may be a human body, or it may be another mammal. The warmth causes the botfly eggs to hatch, and the larvae drop to the surface of the skin. They burrow down into the skin in a few minutes. They don't go very deep. (This might be a good time to tell you that

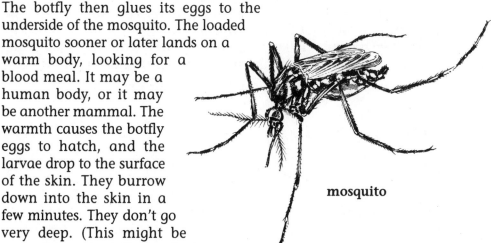

mosquito

mosquitoes in the United States don't carry human botflies. You may be thinking about that mosquito that bit you last week.) Within 2 weeks, a little bump sort of like a boil forms over the place where each fly larva has burrowed. The bump is called a **furuncle.** As the larva inside grows, the furuncle gets bigger. Each furuncle or boil has a small opening at the top. The larva sticks a little tube like a snorkel out of the opening. It's a breathing tube, called a spiracle. Sometimes the larva can be felt moving, especially if its air supply is cut off. This can happen if the host is swimming. And at times the larva can cause pain. But human hosts say that most of the time they don't feel the larva or the furuncle.

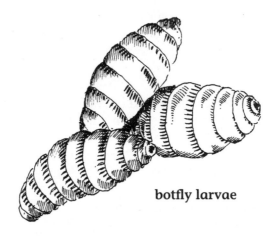

botfly larvae

In the normal course of events, the botfly larva will come out of the bump on its own after about 7 weeks. The exit takes about half an hour. It may involve a couple of false starts before the larva wiggles all the way out. Ideally (for the botfly), the larva will drop to the ground and burrow in the soil, where it sheds its skin again and becomes a pupa. The pupa lies still underground, while inside the pupal skin it changes, or **metamorphoses,** into an adult fly. The fly emerges from the ground and flies off in search of a mate. Unlike most other fly species, the adult botfly cannot eat, because it has no mouthparts. It lives only long enough to mate and, if it is a female, to lay eggs on an insect that will take them to a human host.

Where They Live

Human botflies live only in the forests of Mexico and Central and South America. So if you live somewhere other than these places, you don't need to concern yourself with botflies on your own body.

There are botflies in the United States, but they are not human botflies. They are different species. They can attack livestock and other mammals, but not humans.

Getting Rid of a Botfly Larva

A person who has been infected with a botfly larva has several choices about what to do. One choice is to leave it alone. A larva does very little damage inside its furuncle. It even secretes its own antibiotics and antifungal substances to keep the skin around it healthy. And the hole left behind is not a problem. It will close up and heal when the larva has left to pupate.

Still, many people with a botfly furuncle decide against waiting it out. Some are tempted to squeeze it out or pull it out, but this is not a good solution. The larva has hooks that can dig in and hold on, so it may not come out without tearing into two pieces. Leaving a piece under the skin can cause an infection.

But there are other remedies. Most involve cutting off the air supply to the spiracle. A popular solution is to tie a slab of raw meat tightly over the furuncle. The meat keeps air from getting to the larva. It will often crawl out of the furuncle and on through the meat to the top, where it can get air again. This is what Juan the research assistant had in mind when he suggested the beefsteak.

The safest way to have one removed is by a doctor, although this is not always a choice for people living in the tropics. They may be far away from towns and doctors.

Botflies are not very charming creatures, but they are interesting. They are interesting enough that some researchers have infected themselves on purpose, just to have a genuine botfly furuncle experience!

Madagascan Giant Hissing Cockroaches

That's Strange!

The young male was not happy with the presence of another male in his territory. This was the same guy who'd been hanging around yesterday. And yesterday the young male had had to get rough. He had rammed the newcomer over and over. Then he had put his front legs on the stranger's back and bitten him on the back and legs. Finally that had chased the new guy away.

But now here he was back again, just asking for trouble. Well then, the strong young male was ready to give it to him. First he rammed and butted the newcomer again. But that wasn't enough. So the strong young male stood to his fullest height, ready to unleash his most powerful challenge on the stranger. As the stranger stood there looking blank and dumb, the young male pointed his hind end toward the sky, as high as he could. And then . . . he hissed. The new male cowered, as if to say, "Not *that!*"

But the ruthless young male wasn't finished. He followed the hiss with a violent tail wag. And that did the job. The stranger was whipped, finished. Insulted and beaten, he scuttled away to find a place to hide.

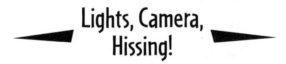

Lights, Camera, Hissing!

When a movie scene calls for roaches, Madagascan giant hissing cockroaches are often used. They don't look exactly like the more familiar roaches, but in movies, it's hard to tell. They're ideal actors because they move slowly, they don't fly, and they're clean and easily handled. If a scene requires smaller roaches, young ones can be used. The roaches that come out of a man's sleeve in the movie *Men in Black* are probably Madagascan hissing cockroaches. Few people would agree to have a hoard of ordinary household roaches stashed up their sleeve. But Madagascan roaches are another matter. If you get to know these hissing roaches, I think you'll agree that they are far removed from their unseemly cousins. Their charms far outweigh their family's bad reputation.

What They Look Like Madagascan giant hissing cockroaches *(Gromphadorina portentosa)* are the only roaches that hiss. They are related to the American cockroach and the German cockroach, two species that have made household pests of themselves. But the Madagascan hissing cockroaches are not household pests. They're not unclean and they don't smell bad, as their unpopular cousins do.

You would probably not recognize them as cockroaches unless you looked at the head and face, which are tucked under and can only be seen from the side or bottom. The face has a distinct cockroach look to it. But from above you can see only the thorax and the abdomen. From above, the Madagascan hissing cockroach looks more like a big pill bug. That's because this type of cockroach has no wings. So you can clearly see the many segments or sections of the abdomen. It is not an unattractive animal. Although the occasional American cockroach on my kitchen counter makes me draw back in disgust, the Madagascan hissing cockroach has no such effect on me.

These Madagascan roaches are about 2½ inches (6 cm) long and 1 inch (2.5 cm) wide when fully grown, which is much bigger than any roach you'll see in your house or outdoors. Younger ones may be much smaller, but look similar. All cockroaches mature by gradual metamorphosis. There is no larva or pupa in this type of development. Rather, the young, called nymphs, look more or less like miniature adults. They become adults gradually, through a series of small growth steps. As a nymph outgrows its exoskeleton, it sheds the tight skin. Underneath, there is a new skin that expands and then hardens. This happens about six times as a young nymph roach becomes an adult.

The adult male Madagascan hissing cockroach looks a little different from the adult female. The male has two big humps on his thorax. Also, his antennae are thicker and more brushy.

Madagascan giant hissing cockroach

Why Do They Do That?

Male Madagascan giant hissing cockroaches are territorial. This means that each adult male spends most of his time in the same area. If another adult male of the same species comes into this area, he will try to chase it away and will behave aggressively. Females wander around more freely and hang out with other adult females. The females are not territorial or aggressive.

How does a male roach get aggressive? If another male comes too close, the two males will first make contact by touching each other with their antennae. Then the male that is feeling territorial will tuck its head under even farther than usual and use the two knobs on the thorax to butt the other one. The attacked animal may give up and flatten himself, as a dog may creep on its belly if you scold it. But if the attacked roach resists the attack, then the territorial one may bite his legs, back, or antennae. Or

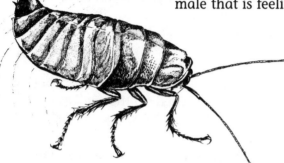

male Madagascan roach, posturing

instead of butting and biting, the territorial one may do something a little more peculiar. He may stand on the tips of his legs, stick his hind end up in the air (called *posturing*), and hiss. The hiss is like the sound you make if you put your top front teeth against your lower lip and blow air out, as in the letter *f*. The cockroach makes this sound by blowing air out of a pair of spiracles, or breathing holes, on the side of the abdomen. Adult insects have several spiracles along the abdomen, one pair per body segment. The two that make noise in hissing cockroaches are different from those that are just for breathing. The territorial roach may also wag his hind end back and forth (called *abdomen thrashing* by roach scientists).

Most roach scientists agree that hissing first developed as a way to frighten predators that might want to eat a roach. Hissing is a warning sound—it makes the predator think the animal might be getting ready to attack. Cats may hiss when they're upset, and are getting ready to scratch or bite. Rattlesnakes also may hiss when they feel threatened and are considering biting. Although some hissing animals may be dangerous, hissing cockroaches are harmless. They don't bite, sting, or spray.

The hissing has taken on new uses in modern Madagascan hissing cockroaches. It became useful for scaring away other roaches as well as predators. The sound probably doesn't actually frighten other male roaches, it just signals to them that the hissing male means business about his territory. It means "Go away. This is my space." It has the same meaning as a male frog's croaking, a male songbird's singing, or a male cricket's chirping. They are all warning other males of the same species to stay away. They also may be saying to females of the same species, "Come on over. I will be your mate." This is true for male Madagascan hissing cockroaches, too. Although in this case, the hiss that is for females (a courtship hiss) is a slightly different sound than the hiss for predators or the hiss for other males. The courtship hiss is somewhat softer and shorter than the other hisses. Its purpose is to attract females and to interest them or prepare them for mating. The loudest roach hiss is the one made when the roaches are disturbed by predators or handled by a person.

Female Madagascan hissing cockroaches hiss when disturbed, but do not make territorial or courtship hisses. Females do have another talent that is perhaps more special and sets them apart from most insect mothers. Most types of insects reproduce by laying eggs and then abandoning the eggs. But female Madagascan cockroaches give birth to live young. They do actually have eggs, but the eggs are held inside the mother's body until they hatch. After the young are born, the mother roach protects them by raising her body and standing over them when they are threatened.

A mother that offers protection to her offspring can't have hundreds of babies at once, as flies and many other insects do. And she doesn't need to. Youngsters that receive some care and protection are more likely to survive. The Madagascan hissing cockroach has only 20 to 40 babies at a time, and at birth they already are about ¼ inch (6 mm) long. At this size, they are already past the most helpless stage of life. So the Madagascan cockroach, with 20 to 40 well-developed and protected young, may be more likely to have babies that survive to adulthood than a mother fly that lays hundreds of eggs and then abandons them.

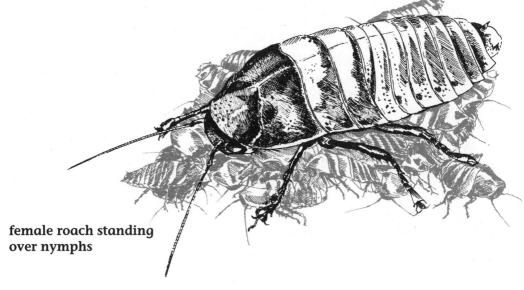

female roach standing over nymphs

Where They Live

Madagascan giant hissing cockroaches are from Madagascar, an island off the southeast coast of Africa. This island has a warm, tropical climate, so the hissing roaches in captivity need temperatures between 65°F and 80°F (18°C and 26°C). Normally, in Madagascar, they live outdoors and eat vegetation and dead material.

Most of the 3,500 species of cockroaches worldwide mind their own business and stay outside. Only a very few species are typically household pests.

What You Can See and Do

You can order Madagascan giant hissing cockroaches from Carolina Biological Supply Company (see "Resources"). Their cockroaches are bred in captivity and are not captured in the wild. So if you buy one, you are not affecting their wild population. These roaches are very easy to keep in captivity. They can live as long as 2 or 3 years. Instructions for their care will come when you order them.

I have three adults that have been with me for about 10 months. They live in a plastic terrarium about 12 inches (30 cm) long, with a plastic snap-on lid. The bottom is covered with chunks of dead wood from a dried-out stump. But pine shavings, bark, or gravel on the floor would be fine, too. If you use pine shavings or gravel, put in some paper towel rolls or egg cartons to serve as hiding places. The roaches prefer to stay out of sight during daylight hours.

When temperatures are warm enough (75°F [23° C] or higher), the cockroaches may breed. One of my females recently gave birth to about 20 or 30 babies. Each is now about ½ inch (12 mm) long. Because of their bare segmented abdomens, they look a lot like pill bugs. The mother stands over them on a piece of dried wood. Not all can fit under her. The others stand in the crevices underneath her piece of wood. Sometimes two or three stand under their dad, although he ignores them entirely.

Madagascan hissing cockroaches do not normally become household pests in Madagascar. However, if a captive Madagascan roach has babies in a cage in your home, it is possible that they can escape into your house and become a problem. So keep a thin cloth over your terrarium, held in place by the terrarium lid. Babies may be able to crawl through the slits in a snap-on lid.

These roaches never bite. To hold a hissing roach, place a finger on each side of the thorax and lift gently. They have hooks on the ends of their legs that may catch on clothes, so move slowly when removing one or you may injure it. Most of the time they are very slow moving. If you put one on your arm, it will generally stay there until you move it.

Army Ants of the Americas

That's Strange!

An army is coming! You can hear them moving through the forest, getting closer. This army is unstoppable. They climb over or under everything in their path. You can even smell them coming. Whew! This army hasn't had a bath in weeks! There is a mad scramble as living things try to get out of the way—running, jumping, flying, bumping into each other. But escape is difficult. The front line of advancing soldiers is almost too broad for fleeing creatures to go around its flanks. And behind the front line are hordes of marching bodies. Panic is in the air. Some of those in the army's path try to quickly find cover. Some even try to dig holes to hide in. But it's of little use. Almost all will be found.

For this is an army of ants and they can crawl up every twig, explore every burrow, find every creature. The army ants will cut up and carry away with them all small creatures that they find. Later they will feed the meat to their queen and her young.

What They Look Like The swarm of army ants is composed of worker ants and soldier ants. All ant colonies have workers of one kind or another. And most have soldiers. In general, workers collect food and they feed and care for the other members of the colony. Worker ants do most of the work that has to be done. Soldiers defend and protect the workers and the rest of the colony.

When army ants are on the march, soldiers lead the way with workers close behind. The soldiers are bigger than the workers. They are well equipped for their jobs, and look more threatening than your average ant. Their jaws are longer and heavier than most ant jaws, and are shaped like the a J or a fishhook. The curved tips point toward each other like ice tongs. These sharp jaws, or mandibles, work very well at cutting up other insects.

army ants

There are lots of different species of army ants, and their size varies. Many are about ⅓ to ½ inch (6 to 12 mm) in length. A common species of army ant in Central America is *Eciton burchelli.* The ants of this species have bodies that are black and orangish red.

Why Do They Do That?

Army ants are predators—they eat other animals. This is not unusual. Many ants are predators. But army ants travel in huge groups looking for prey, and that *is* unusual. They are very much like a real army, with the fiercest soldiers in front. They attack and cut up almost every small creature they come across, especially insects. As a **swarm** (large group) of army ants approaches, you can hear the rustling sounds of insects and other animals trying to flee. Large animals such as adult mammals or adult birds or humans have no trouble moving to safety. But baby birds or sleeping frogs may be killed, first stung by poisonous stingers and then cut into pieces.

The many different species of army ants differ in their hunting habits. Some hunt by night, some by day. The formation of the hunting parties

◣ Opportunists and Impostors ◢

Flying insects have a chance of escaping army ants. But they are likely to be eaten by the flocks of antbirds that fly behind the swarm. Many species of tropical antbirds make their living almost entirely by following army ants. They grab not only flying insects, but other small creatures here and there that are overlooked by the ants.

Some types of beetles and wasps may travel along with army ants, too, taking advantage of all the food that gets stirred up. Some are protected from the ants by their smell, which mimics the smell of the ants. Ants rely on this smell to identify their family members. Others with the same smell are accepted as family, too.

**antbird with swarm
of army ants**

can differ, too. Some begin as a column in early morning, then slowly spread out into a fan shape. A swarm like this can stretch 100 feet (30 m) across at the widest point. Army ants that hunt in this formation are called swarm raiders. As they sweep over an area, they pick it clean. This can be an advantage to plants, because the ants eat all the insects that were eating the plants. It can also be an advantage to insects that taste bad and don't get eaten. Now they will have less competition for food and nest sites.

Some types of army ants always hunt in columns. They are called column raiders. I first encountered column raiders on a dark night in a rain forest of Central America. I was walking a trail through the forest, heading back to the clearing where I was staying.

People who live there will tell you to always use a powerful flashlight at night in the rain forest, to light up the trail. Otherwise you might step on a dangerous snake, like the very poisonous fer-de-lance or a coral snake. I wasn't paying much attention to the trail itself, but I did have a flashlight—a dim one. Suddenly I was aware that something very long was moving down the trail beside me, along the edge where the trail met the forest. A huge snake! I jumped, and yelled to my three companions behind me. When we all put our flashlights on it, we realized it wasn't a snake at all. It was a column of army ants! The column stretched as far as we could see ahead of us and behind us. Oddly, the ants were moving in both directions in the same column, although most of them seemed to be headed in the direction we had just come from. Right beside us was a big knot in the column, a ball of ants the size of a baseball. The ants in the knot were swarming all over each other. We tried to pull the ball apart with a couple of sticks, and we saw that even the ants inside the ball were crawling and squirming. There was a small roach or beetle in the middle, who seemed to be the object of their attentions.

Army ants travel through the jungle at a fairly rapid rate—up to about 1 foot (.3 m or 30 cm) per minute. They can cover a lot of ground in a day. As they sweep through, they eat just about every creature that is edible. They have to keep moving and may not return for many months to an area they've just raided.

Well, this wandering life is fine for the workers and soldiers, who are long-legged and fast walkers. But what about the queen and the eggs, larvae, and pupae of the colony? Your average ant that is not an army ant lives in a burrow with lots of rooms underground, or perhaps in a burrow in a tree, stump, or log. The queen and the immature stages (larvae and so on) stay inside the nest all the time. The soldiers protect them. The workers go out and find food, bring it back, and feed everyone else. They take care of the queen and the young, who stay home.

This works very well, unless you're a traveling ant. Then you have a problem. But army ants have come up with a solution to this tricky problem of how to care for queen and young when you're an ant on the go. The army ant queen and the young all live in a temporary camp or **bivouac**. Every night the army ant workers form walls around the queen and all the nonadult ants. They make the walls with their bodies! The workers hook their bodies together using the claws on the ends of their legs. The queen and the young are safe inside the ant walls. At dawn the walls break up and turn back into active ants.

wall of an army ant bivouac

The queens of ordinary ants may lay eggs almost constantly. But army ant queens are different. The army ant queen lays eggs for several weeks, then stops for several weeks, then begins again. During the time she is laying eggs, she is very big—the size of a baby mouse—and cannot walk. So for a few weeks she doesn't travel at all. The workers and soldiers that forage (search for food) must come back to her every night in the same spot. Some of the workers and soldiers stay with her all day to protect her and to care for her and her eggs. This phase of colony life is called the stationary phase. *Stationary* means "staying in one place."

Then the queen stops laying eggs. Her body shrinks and she can walk. Now, every night, when the workers and soldiers come home from a long day of raiding, they pack up camp and move to a new site. The queen may walk on her own to the new camp, surrounded by workers, or the workers may carry her. The workers also carry the nonadult ants. This phase of their life is called the migratory phase. *Migratory* means "moving from one place to another."

The eggs that the queen laid during the stationary phase begin to hatch as the stationary phase ends. They hatch into larvae that are very hungry. Suddenly the food demands of the colony are much greater. So now the workers and soldiers must cover new ground on their raids. They must find a lot of food to bring home to those hungry larvae. And fortunately, they are able to move to new areas of forest now, because the queen has gotten smaller and is walking or is easier to carry.

About the time the larvae reach their maximum size and turn into pupae (which don't eat), the queen swells up and starts laying eggs again. And again the workers and soldiers must settle for foraging in a limited area for several weeks. But that's OK. There are no larvae to feed, only the queen and the workers themselves.

◄ In Stitches ►

Here's another strange thing about army ants. They have a medical use! Their jaws can be used instead of stitches for closing a cut in the skin! Some of the Indian tribes of South America are too far into the jungle to get modern medical care. So if there is a bad cut, a soldier army ant is held over the wound and squeezed in a way that makes it close its jaws. Its tonglike jaws pull the sides of the cut together and hold it, in the same way that stitches would. Then the ant's body is pinched off, leaving only the head with closed jaws. Several soldiers are needed to hold a cut together—each head is like one stitch. As the wound heals, the heads fall off!

Where They Live

Army ants live in forests in tropical areas of North and South America. Their range can also extend up into the southwestern United States.

Driver ants live in Africa. They are even more fierce than the army ants of the Americas. Their swarming, predatory behavior is similar, but they are a different genus and species from the American varieties.

What You Can See and Do

If you are fortunate enough to see a swarm or column of army ants, watch them carefully. You will probably notice that their movements seem disorganized and chaotic. They crawl over one another, going in different directions at the same time. But they can cooperate in attacking an insect that they stumble across. Some may pull the insect's legs out straight, while others use their mandibles to clip the insect into pieces at the narrow parts and joints.

Listen and look for antbirds following the swarm. They will be swooping after insects that may have hopped or flown out of the way.

Sniff the air. Army ants secrete chemicals to communicate with each other. Sometimes people can smell them. The queen in particular secretes powerful scents that attract the workers, and cause them to stay near her and care for her. Odors that insects and other animals produce to communicate with each other are called **pheromones**. Some people say that army ant pheromones smell musky; others say they stink.

If you stand in their way, army ants will crawl up your pants legs. They may sting you, but the sting is not that bad. You can knock them off and move away. They are not like the driver ants of Africa, which can kill and eat a horse if it is tied up!

Bombardier Beetles

That's Strange!

Imagine that you are a small, hungry mouse. Night has finally fallen, and you're ready to go out looking for food. You leave the toolshed where your home is and move slowly out into the yard. You know it's risky being in the open yard, but all the seeds under the bird feeder are irresistible. You make it to the feeder and crouch under it for a couple of minutes, gathering seeds. Then the back door opens and you dash wildly back to the shed. So much for the bird feeder. What else can you try tonight?

You decide to look in the woods behind the shed. You leave the shed again cautiously, and walk along the edge of it very quietly. You move into the woods slowly and begin looking among the leaves for something edible. Ah, a little worm! That's tasty—down it goes. You root around for several more minutes and finally see a little black and tan beetle scurrying for cover. Your mouth begins to water as you reach for the little insect. Then *pop!* What was that? Ouch! Your eyes are burning and stinging! You stop to rub your eyes. When you open them again, the little beetle is gone. What was it? A type of ground beetle called a bombardier beetle. And you've been bombed.

bombardier beetle

What They Look Like

Bombardier beetles are in the genus *Brachinus*. This genus is part of the beetle family called Carabidae. It is one of the largest families of beetles, with over 3,000 species in North America. Many of the species are very common and easy to find, at least in the eastern United States.

Most beetles in the family Carabidae (called carabids) are black and shiny. Many have very fine grooves running the length of their elytra (wing covers). They all have long thin legs and are fast runners. Carabids also have narrow heads. The head is nearly always narrower than the prothorax (the part of the thorax just behind the head), which is not true for all beetles.

Some carabid species are brightly colored. One bright carabid is the fiery searcher. Its elytra are greenish with red edges. The fiery searcher eats caterpillars and sometimes climbs trees in search of them.

Another carabid that's brightly colored is the bombardier beetle. The head and prothorax of the bombardier are yellow or orangish tan and very narrow. The elytra are blue to dark blue to black. Its legs are reddish yellow.

All members of the family Carabidae can be called ground beetles, because most spend their entire lives on the ground. As a group, ground beetles range in length from $\frac{1}{8}$ to $1\frac{3}{8}$ inches (3 to 35 mm). But most are between $\frac{3}{16}$ and $\frac{3}{4}$ inch (5 and 20 mm). The 40 species of bombardier beetles in the genus *Brachinus* range in length from $\frac{1}{8}$ to $\frac{5}{8}$ inch (4 to 15 mm).

Why Do They Do That?

Like most ground beetles, bombardier beetles are active hunters. They are predators, which means that they catch living animals and eat them. Since ground beetles are small, their prey are also very small. Most predators have some sort of special feature that helps them to catch prey. A praying mantis has spiky, grasping forearms. A spider has a web. A snake has sharp, pointed teeth and a low profile. A cheetah has speed, and so does a ground beetle. With their long thin legs, ground beetles are very fast runners. Although ground beetles in general are fairly easy to find, they are hard to catch. If you turn over a board and see one, it will be gone within 2 or 3 seconds.

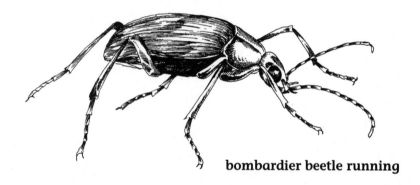

bombardier beetle running

Their speed helps them catch prey, and it also helps them avoid becoming prey for bigger predators. Beetles are known for having a broad range of defenses. Their hard, shiny elytra are one of their most effective defenses. Many ladybug beetles and other dome-shaped beetles, when attacked, will clamp down onto whatever they're standing on, pulling in their legs and antennae for safety. The sharp jaws of a small predator can't grip the slippery elytra.

Staying hidden is also a very effective defense, and ground beetles are very good at that. They hide under logs and other objects during the day. Only when it's dark and safe do they come out to look for food.

Lots of beetles use different forms of chemical defense. Some ladybug beetles release a sticky, distasteful fluid from their leg joints if a predator bothers them. Blister beetles have a substance called cantharadin in their elytra and body fluids that can cause blisters on the skin of people or predators. The larvae of Mexican bean beetles are covered with thin, branched spines that contain a sticky liquid. When something touches them, the spines break and the sticky liquid leaks out. If it gets on the jaws of a predator, it gums them up and may not come off. These are all chemical defenses. They are all examples of beetles protecting themselves by using some type of chemical that is either unpleasant or harmful.

The master of chemical defense in the beetle world is probably the bombardier beetle. It does more than just ooze foul liquids, more than just taste bad. Many of these other beetles with chemical defenses wait passively until they are grabbed or bitten to deliver the bad chemicals. The bombardier beetle is more aggressive. It fires the bad chemicals at the enemy like a cannon. *Bombardier* means "one who bombs." When another creature comes too close to the bombardier beetle, the beetle raises or lowers its hind end and aims it at the offender. With a *pop!* it releases a spray of toxic chemicals. The spray comes out at the temperature of boiling water! Toads have been observed trying to eat these beetles. After being sprayed in the mouth, the toads appear to be in pain—opening their mouths wide and rubbing their tongues against the ground.

bombardier beetle spraying at a predator

But the spray of toxic chemicals doesn't always help the beetle. Certain mice have learned to pick up a bombardier beetle and ram its hind end into the ground while it shoots out its nasty spray. After the chemicals are all gone, the mouse eats the beetle—head first.

Where They Live

Bombardier beetles live throughout the United States and southern Canada. But they are not easy to find because they live only on floodplains of rivers or lakes, in areas where temporary ponds form after rainstorms or floods.

Like almost all ground beetles, they live on the ground. During the day, they stay hidden under leaves, logs, stones, and other cover. At night they go out to look on the ground for food.

An Explosive Mixture

How do bombardier beetles spray out chemicals that are the temperature of boiling water? The popping sound and the release of the toxic spray are the result of an explosion that occurs inside the body of the bombardier beetle. A scientist named Thomas Eisner described how the bombardier produces the explosion when needed. The beetle has an organ with two special storage chambers. One contains hydrogen peroxide. (You may have some hydrogen peroxide in your bathroom closet. It is often used to clean cuts and scrapes.) The bombardier's other storage chamber contains chemicals called hydroquinones. When the beetle needs a defensive spray, the substances in these two chambers are thrown together into a third chamber with certain enzymes. An explosive reaction occurs that produces water, oxygen, heat, and some toxic, irritating substances called benzoquinones. The explo-

sion expels the mixture with force. The heat causes some of it to vaporize, or turn into a gaseous cloud. This hot spray and cloud are very irritating to skin and eyes. They can cause a blister or a painful red spot on human skin.

Bombardier beetles don't use all of the chemicals in one explosion. They can fire off many rounds in rapid succession if necessary. Or they can fire a couple, and save the rest for later. If they do use up all the chemicals in their storage chambers at one time, then they are unable to fire again until they can manufacture more.

The tip of a bombardier beetle's abdomen works a little bit like the nozzle of a fire hose. It can swivel around, to aim the spray either forward, backward, or to either side.

Why do the chemicals not hurt the bombardier beetle? After being mixed, the chemicals are inside the beetle's body for only a fraction of a second before they are expelled.

What You Can See and Do

If you happen to live near a floodplain, look under logs, boards, and other debris during the day. If you find a bombardier beetle, move quickly to trap it against the ground with the open end of a clean cup. Then slowly slide a piece of cardboard under the cup, so that the beetle is enclosed in the cup. Turn cup and cardboard right side up and the beetle should be in the cup. You can probably get it to fire its weapon by putting something into the cup. If you are very quiet, you can hear the popping sound. You may be able to see a little cloud coming out of its hind end.

After you've seen what the bombardier can do, let it go where you found it. Don't exhaust the little beetle's weapon by making it fire several times. If you do, the beetle will be without its main defense when you let it go.

Honey Ants

That's Strange!

Picture in your mind a golden ball, almost marble-size. Sunlight passing through it gives it a warm, amber glow. This ball is full of a sweet liquid that's like honey. Doesn't it sound tasty—like a special kind of candy? Some think so. The Aborigines of Australia have eaten them. But you might not find them very appetizing—because the ball has a thorax, a head, and six legs. It is alive and kicking. The golden ball is the abdomen of an insect called a honey ant, or honeypot ant. The ant's body is used as a storage chamber for this sweet liquid food. It's a little honey jar that wiggles and bites.

What They Look Like

Because there are so many different species of honey ants, their appearance varies. Many are black or reddish brown or both. Their size varies, too, but workers tend to be less than ½ inch (13 mm) in length. The swollen abdomens, full of "honey," can be as large as a blueberry!

replete honey ant,
full of stored food

Why Do They Do That?

Ants are known for their willingness to give their lives for the sake of the queen, who is their mother, and for the other workers, who are their sisters. But this is truly a story of self-sacrifice. Honey ants usually live in desert areas where droughts occur. During very dry periods, the ant colony may not be able to find enough to eat. Just as squirrels store nuts for winter, honey ants need to store food for times of need. But much of their diet is liquid. They eat honeydew from aphids, and nectar from a type of **gall** (an odd growth) on oak trees—oh yes, and chewed-up termites. Where can you store stuff like that? You can't just spit out droplets on the ground and expect them to stay there for months until you need them. You need some kind of container.

Well, now that you mention it, all worker ants have a special stomach just for storing small amounts of food—the amount you might want to feed a few larvae or a few other workers. This special storage stomach is called a crop. A worker that has been out foraging can bring home a little food for her sisters and coworkers.

What if a foraging ant could cram just a little bit more than usual into her sister's crop, and then make her sit still and hold it for months? That's sort of what honey ants do.

Over evolutionary time, the crops of the honey ant workers have become more and more elastic, until their crops now can accept huge quantities of stored food. The process works like this. Some of the workers go out and collect food—the honeydew from aphids, the nectar from galls, water, and chewed-up termites. They bring this stuff back to the colony. Then they go up to one of the large workers who has stayed home and who is willing to accept donations. The food is transferred to the larger worker by mouth. A worker who accepts donated food is called a **replete**, or honeypot. She only consumes a part of it for her own nourishment. The rest is just stored in her crop. She may accept donations from hundreds of workers. As her crop and her abdomen swell, the segments of her abdomen come apart, so that the thin, stretchy membrane between them shows. The membrane is almost clear, and you can see the golden liquid inside.

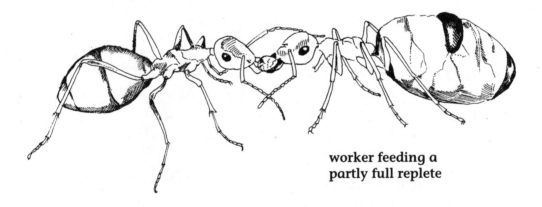

**worker feeding a
partly full replete**

When the replete is as full as she can get, she climbs to the ceiling of an underground chamber and hangs there by her claws. She hangs with hundreds of other repletes until they are needed. If a shortage of food occurs, the workers will come to the replete and draw from her mouth a drop of food.

If the replete's stored liquid is all used up, she cannot regain her original slim shape. She is too stretched out, and usually dies.

repletes hanging from chamber ceiling

◀ My Colony's Bigger Than Your Colony ▶

While he was studying honey ants in Arizona, a scientist named Bert Hölldobler noticed that at times the workers aboveground behaved very oddly. They stretched their legs out as though walking on stilts. While prancing around in this odd way, one worker would approach another worker from a neighboring honey ant colony. The two workers would stand tall, even standing on pebbles to make themselves look taller. Then they would walk in circles around each other, kicking at each other and finally pushing and shoving. After a while, one of the ants would break away and go find someone else to spar with. Hölldobler took a lot of films of the ants interacting this way. It was sort of like a tournament between colonies, a contest that could last for weeks.

Hölldobler learned that the ants were really sizing each other up. After a while, the ants could tell which colony had more workers. If one colony had a lot more, then the workers of that colony might decide to stop this polite little dance and go for it—kill the workers of the smaller colony and charge into their underground nest. There they would kill the queen and steal the colony's repletes for their stored food. They would steal the immature ants, too, to raise as slaves. When grown, the slaves would be forced to work for the raiding colony.

Where They Live

Honey ants live on three different continents: Africa, Australia, and North America. They belong to several different genera in two different subfamilies.

Many honey ants live in dry, desert areas where droughts occur. Others live in damper soil. The honey ants in Arizona, in the United States, have been studied very carefully by scientists. They belong to the genus *Myrmecocystus*. Much of the information in this chapter is based on the results of these studies.

Honey ants live in underground burrows with many chambers. The burrows are often in very hard, dry soil. Some of the chambers may be as deep as 16 feet (5m)! This may be for protection of the repletes.

What You Can See and Do

Honey ants may be hard for nonscientists to identify, because the swollen ones (the repletes) are not visible aboveground. If you're lucky enough to find a couple of colonies, you can look for the contests between workers.

Tsetse Flies

That's Strange!

A few hundred years ago, all of Africa was a place where wild animals roamed freely. There were no paved highways, no cars, no industry, no parking lots to destroy their habitats. The humans who lived there lived in harmony with nature, taking only what they needed.

Then the Europeans came. They changed everything they touched. They built cities, paved animal habitats, altered animal populations.

There was one creature that slowed the spread of the Europeans. It slowed them by killing them and their horses and cattle. The actual killer was a disease called sleeping sickness. This disease begins with fever and swollen glands. It ends months or years later with a fatal swelling of the brain. The Europeans didn't understand what caused the illness.

Sleeping sickness is carried and delivered by a fly called the tsetse fly. Dr. David Livingstone, a famous explorer of Africa, guessed that the tsetse fly causes illness and death by injecting a toxin into its victims. But he was wrong. The tsetse fly does inject something into its victims, but it isn't a toxin.

adult tsetse fly

What They Look Like

There are 22 species of tsetse flies, although only 5 of the species are responsible for causing most cases of sleeping sickness.

Tsetse flies look very much like houseflies, with two large compound eyes on the front of the face. But tsetse flies are larger—$\frac{1}{3}$ to $\frac{1}{2}$ inch (7 to 13 mm) in length. And their color ranges from yellowish to dark brown, while houseflies are grayish. Tsetse flies also differ from houseflies in the way the wings are folded when not in use. The wings of tsetse flies cross each other on the back like a pair of scissors. Each wing of a housefly lies straight back when not in use.

sawlike structure for tearing skin

Both tsetse flies and houseflies suck up their food. But the structures for sucking are different. A housefly's mouthpart looks rather like an elephant's trunk—long and flexible. Put a touch of sugar water on a table in front of a housefly sometime. You will see it slurp up the liquid with his long proboscis. The proboscis of a tsetse fly is different. It looks like a hypodermic needle that has two channels. And it punctures like a hypodermic needle.

Tsetse flies also have an odd sawlike structure on the front of the face. It sticks straight forward (parallel to the ground) from the area near the proboscis.

Why Do They Do That?

Tsetse flies feed by sucking blood from humans and other animals. When a tsetse lands on a victim, it uses the sawlike structure on the front of the face to tear through the skin, causing bleeding. The tsetse fly then uses its proboscis for eating. One channel in the proboscis delivers saliva to the small pool of blood on the victim's skin. In the saliva is a substance that keeps the blood from clotting, or forming a scab. The other channel in the proboscis sucks up the blood.

A mosquito is another type of fly that sucks blood. It doesn't look like a housefly, but a mosquito is a fly. It's in the order Diptera, which includes all the flies and no other insects.

When mosquitoes feed, they also inject a substance to prevent clotting. This substance is what causes the allergic reaction that produces an itchy bump. But mosquitoes have no sawlike structure to tear the skin and cause bleeding. Instead, a mosquito sticks its proboscis straight into a capillary (a tiny blood vessel) to get blood. So the mosquito makes only a tiny puncture.

What is it about tsetse flies that causes sleeping sickness? These flies can have tiny one-celled parasites called trypanosomes living inside their proboscis, salivary glands, or other organs. When the tsetse fly bites a human or animal, the trypanosomes pass into the victim. Then these tiny one-celled parasites multiply inside the victim's body and cause sleeping sickness.

A tsetse fly that doesn't have trypanosomes can get them from biting an infected animal or human. And then it passes on the trypanosomes to the next animal or human that it bites.

Trypanosomes are parasites, and so are tsetse flies. Parasites are animals that feed on other living animals without killing them, at least not right away. Some parasites, such as trypanosomes and tapeworms, live inside the body of their host. (The host is the animal that a parasite feeds on.) Other parasites live on the outside of the host's body, but stay there all the time, as fleas, ticks, and lice do. There are still other parasites that don't live in or on their host. Like mosquitoes and tsetse flies, they only drop in to visit the host for meals. Eat and run.

How do parasites locate their hosts? A lot of parasites that feed on mammals can find their hosts by sensing the host's body heat. Ticks wait on a bush or branch, and drop onto a mammal when they sense the warm body underneath them. Mosquitoes also find their victims by sensing body heat. But tsetse flies use their sight to locate victims. In fact, researchers have discovered that some tsetse flies can be fooled by anything that looks like a big, dark mammal. Some will fly to a car moving slowly across the African plain, if the car is covered with a gray blanket. Flies that come to the car can be caught and killed, so this is a handy trick in trying to control the flies.

Probably the very strangest thing about tsetse flies is the way that they reproduce. The usual method of reproduction in insects is to lay eggs and leave them to hatch and grow up on their own. Some mother insects, such as stinkbugs, earwigs, dung beetles, burying beetles, and hissing cockroaches, provide a certain amount of protection or care for their young. But the great majority of insect species provide no parental care at all.

Mother insects that protect or care for their eggs or young in some way usually have fewer young. Their young are usually more likely to survive than those that are ignored, so they don't need to have as many. Insects and other animals that ignore their young must produce large numbers, because most of them won't survive.

One way to protect offspring is to keep the eggs inside the mother's body until after they've hatched. Eggs are completely helpless. They can't move to escape a predator or other threat. So an egg or a very young larva would really be much safer inside the mother's body. Mother is bigger and can move to escape danger, or maybe even defend herself.

Tsetse flies have adopted this method of child rearing. A mother tsetse fly keeps her eggs inside her body. But I should say "egg" because she keeps only one at a time. Many insects lay hundreds of eggs at a time, but Mama Tsetse has only one at a time. She keeps this single egg in her body. She keeps it inside not only until it hatches, but all through the larval stage—until it's ready to pupate! This is much more like reproduction in elephants, humans, or cows than in other insects! So Mother Tsetse gives birth to a single full-grown larva.

This is a great advantage to the larva, and greatly improves its chances of survival. The larva has such a good chance of survival that, over her entire lifetime, the average mother tsetse fly produces probably only two or three larvae. If she has a very long life, she could produce as many as 10 or 12 larvae. It is very, very unusual for an insect to produce such a low number of offspring. But obviously this strategy works well for the tsetse fly, because we still have plenty of tsetse flies around.

There's one problem with the tsetse fly larva growing up inside its mother's body. If it grows, it must be eating. But what is it eating? Baby mammals spend a lot of time growing inside their mothers. How have they solved this problem? Most mammals have a placenta, a structure that connects the baby to the mother's blood through the baby's umbilical cord. Then the baby can get all the nutrients that are in the mother's blood from the meals that she has eaten. Other mammals, the marsupials, do it a little differently. A baby marsupial is born at a very early and very small stage of development. It crawls up to the mother's pouch and stays inside there for weeks or months. It feeds on milk from a nipple inside the pouch. When the baby is big enough to move around by itself some, it begins to come out of the pouch for short periods. As it grows up, it stays out more and more.

Amazingly, the mother tsetse fly nourishes her larva in much the same way. The larva stays inside the mom in a structure similar to a uterus—a sort of internal sac for the baby. And the larva feeds on a white liquid that comes from a sort of nipple inside the "uterus." It's not really milk, but it is a lot like milk.

When the larva has grown and is ready to become an adult, it is "born," or pushed out of the mother. It burrows into the ground and right away becomes a pupa. During this pupal stage underground, its body metamorphoses into the body of an adult tsetse fly. After about 2 weeks, the new adult is fully formed and out it comes, ready to look for a mate.

Where They Live

Tsetse flies are found in three areas that cut across the middle of Africa. Some places in these areas are nearly uninhabitable because of tsetse flies and sleeping sickness.

◄ Flies That Don't Fly ►

The tsetse fly has a relative called the sheep ked, whose lifestyle is very similar to the tsetse fly's. The sheep ked is another parasitic fly. It sucks blood from sheep. Its mouthparts are very similar to the tsetse fly's. And like the tsetse fly, the mother sheep ked produces one larva at a time and keeps it inside her body until it is ready to metamorphose into an adult. But unlike the tsetse fly, the sheep ked has become a full-time parasite. It not only feeds on sheep, it lives its entire life on sheep. When the grown sheep ked larva comes out of its mother, the mother glues it to the base of a sheep hair. And there it pupates, inside the larval skin. The adult that emerges a couple of weeks later is fully adapted to life on a sheep. It has no wings! It doesn't need wings—it's not going anywhere. And wings would only get in the way as the fly walks around among the sheep hairs. Sheep keds look a lot like ticks. Their abdomens swell with blood when they're feeding. But they're not ticks, they're flies. (Ticks are not even insects. They're more closely related to spiders.)

sheep ked

Resources

Ordering Insects

To order Madagascan giant hissing cockroaches, giant water bugs, and other insects, contact:

> Carolina Biological Supply Company
> 2700 York Road
> Burlington, NC 27215
> 800 584-0381

> Ward's Biology
> PO Box 92912
> Rochester, NY 14692
> 800 962-2660

Insect Zoos

> The Butterfly Pavilion
> The Nature Museum
> Charlotte, NC
> 704 372-6261, ext. 605

> Butterfly Pavilion and Insect Center
> Westminster, CO
> 303 469-5441

> Cincinnati Zoo's World of Insects
> Cincinnati, OH
> 800 94-HIPPO

> Insect Zoo
> Oregon Zoo
> Portland, OR
> www.zooregon.org

> The Insect Zoo at Iowa State University
> Science II Building and Insectary Building
> Iowa State University
> Ames, IA
> www.ent.iastate.edu/zoo/enhanced4.html

> The Insectarium at Steve's Bug-Off Exterminating Company
> Philadelphia, PA
> (215) 338-3000

The Ohio State University Insect Collection
Columbus, OH
iris.biosci.ohio-state.edu/inscoll.html

O. Orkin Insect Zoo
Smithsonian Institution National Museum of Natural History
Washington, DC
www.orkin.com/html/insect_zoo.html

Ralph M. Parsons Insect Zoo
Natural History Museum of Los Angeles County
Los Angeles, CA
213 744-DINO

San Francisco Zoo's Insect Zoo
San Francisco, CA
415 753-7080

Victoria Bug Zoo
Victoria, BC
Canada
250 384-BUGS
www.bugzoo.bc.ca

Further Reading

Borror, Donald, and Richard White. *A Field Guide to the Insects.* Peterson Field Guide series. Boston: Houghton Mifflin, 1970. If I could have only one field guide to the insects, I would choose this one.

Facklam, Margery. *The Big Bug Book.* New York: Little, Brown, 1994. Large beautiful illustrations, short text.

Forsyth, Adrian, and Ken Miyata. *Tropical Nature: Life and Death in the Rain Forests of Central and South America.* New York: Touchstone, Simon & Schuster, 1984. For advanced readers or adults. This is one of my favorite books. It has chapters on army ants, leaf-cutter ants, botflies, and many other fascinating rainforest creatures.

George, Jean Craighead. *One Day in the Tropical Rainforest.* New York: HarperCollins, 1995. The story of a Venezuelan Indian boy's efforts to protect the rain forest and its wildlife.

Goodman, Susan E. *Adventures in the Amazon Rain Forest (Ultimate Field Trip, No. 1).* New York: Aladdin Paperbacks, 1999. A photo essay about a field trip of middle school students from Michigan to the rain forest of Peru.

Goor, Ron, and Nancy Goor. *Insect Metamorphosis.* New York: Macmillan, 1990. This book has beautiful color photographs of an assortment of insects in various stages of their life cycles.

Greenbacker, Liz. *Bugs: Stingers, Suckers, Sweeties, Swingers.* New York: Franklin Watts, 1993. Discusses insects in and around homes.

Hoyt, Erich. *The Earth Dwellers: Adventures in the Land of Ants.* New York: Touchstone, Simon & Schuster, 1996. An interesting book for advanced readers or adults, written in a relaxed style. It describes in detail the natural history of army ants and leaf-cutting ants in the tropics of Central America, specifically at La Selva Biological Station in Costa Rica. It relates the interactions of these ants and other ants with other animals in the forest and with researchers.

Kneidel, Sally. *Classroom Critters and the Scientific Method.* Golden, CO: Fulcrum, 1999. Describes experiments and science projects you can do with gerbils, mice, goldfish, guppies, hamsters, lizards, kittens, and puppies.

———. *Creepy Crawlies and the Scientific Method.* Golden, CO: Fulcrum, 1983. Describes activities and science projects you can do with some common insects, tadpoles, toads, earthworms, and other animals.

———. *Pet Bugs.* New York: Wiley, 1994. This book has 26 chapters on 26 insects, describing how to find them, how to catch them, how to take care of them in captivity, and how to observe their most interesting behaviors.

———. *More Pet Bugs.* New York: Wiley, 1999. The same format as *Pet Bugs,* but with 23 entirely different types of bugs.

———. *Slugs, Bugs, and Salamanders: Discovering Animals in Your Garden.* Golden, CO: Fulcrum, 1997. Experiments that reveal the connections between common garden plants, the pests that eat them, and predators that eat the pests. Includes a garden map and suggests plants that children will enjoy interacting with.

Lavies, Bianco. *Backyard Hunter, Praying Mantis.* New York: Dutton, 1990. Impressive color photographs of mantises in every stage of life, including hatching from the egg case.

———. *Compost Critters.* New York: Dutton, 1993. Most of the book is about the animals that live in and feed on compost and how they break it down.

———. *Monarch Butterflies, Mysterious Travelers.* New York: Dutton, 1992. Discusses monarchs and their migration.

Leahy, Christopher. *Peterson First Guide to Insects of North America.* Boston: Houghton Mifflin, 1987. A much more limited insect field guide than the standard Peterson guide by Borror and White described earlier. But it's easier for children to use and fits easily into a shirt pocket.

Lovett, Sarah. *Extremely Weird Insects.* Santa Fe, NM: John Muir, 1996. Covers 22 odd insects with color photographs.

Milne, Lorus, and Margery Milne. *The Audubon Society Field Guide to North American Insects and Spiders.* New York: Knopf, 1990. The color photographs make this book fun for browsing. Only relatively common species are illustrated.

VanCleave, Janice. *Insects and Spiders: Mind-Boggling Experiments You Can Turn into Science Fair Projects.* New York: Wiley, 1998. Lots of interesting information and activities.

Waldbauer, Gilbert. *Insects through the Seasons.* Cambridge, MA: Harvard University Press, 1996. Delightful reading for the advanced reader or adult. Describes in a casual and very readable style the life histories of many out-of-the-ordinary insects.

White, Richard. *A Field Guide to the Beetles.* Peterson Field Guide series. Boston: Houghton Mifflin, 1983. A thorough enough coverage of beetles, with much more information on this group than you'll find in an insect field guide.

Zim, Herbert, and Clarence Cottom. *Insects.* Racine, WI: Golden Press of Racine, 1987. A simpler guide than the standard Peterson guide, and more usable for children. Its coverage, though, is limited.

Glossary

abdomen   The rear section of an insect's body, located behind the thorax.

Batesian mimicry   See **mimicry.**

bivouac   A temporary encampment with little or no shelter.

camouflage   The shape, color, pattern, or behavior of an animal that conceals it by making it appear to be part of its natural surroundings.

caterpillar   The long wormlike larva of a butterfly or moth.

chemical defense   Protecting oneself from predators by the use of a chemical, such as tasting bad, smelling bad, being toxic, spraying chemicals, oozing chemicals, or in any way containing or releasing a substance offensive to predators.

chrysalis   A protective capsule enclosing a pupa as it transforms from larva to adult, often specifically used to mean the casing of tough skin around a butterfly pupa.

class   A major category of biological classification, ranking above an order and below a phylum, and composed of related orders.

cocoon   An envelope, often made of silk, that a moth larva forms around itself and in which it passes the pupal stage.

cold-blooded   Having a body temperature that varies with the temperature of the surroundings.

colony   A group of creatures of the same species that live together.

complete metamorphosis   See **metamorphosis.**

compound eye   An eye (as of an insect) made up of many separate visual units.

courtship   The behavior of mature male and female animals that signals to one another their species' identity and their mating readiness. Within a given species, courtship behaviors are usually predictable and consistent.

crop   An organ of the digestive tract specifically used for storing food.

diapause   A temporary interruption in the development of insect eggs or larvae, usually with a dormant period.

division of labor   The partitioning of jobs within a colony so that each individual does only one of the many tasks required to keep the colony going.

dormant   In a resting state in which some bodily processes are slowed down.

elytra (singular **elytron**)   The thickened, hard, or leathery first pair of wings that function as covers for the second pair of wings in beetles, earwigs, and a few other insects. They form the hard back of beetles at rest.

evolution   Change in the characteristics of a species over time, in response to environmental changes. Evolution incolves the gradual splitting of one species into two or more species.

evolutionary   Having to do with evolution.

evolve   To change or develop gradually over thousands or millions of years, in response to environmental pressures.

exoskeleton   The hard outer covering of an insect's body.

family   A category of biological classification ranking above a genus and below an order, and usually composed of several genera.

forage   To search for food.

furuncle   A large bump on the skin sometimes caused by a parasite under the skin.

gall   An abnormal swelling of plant tissue caused by insects or disease.

genus (plural **genera**)   A category of biological classification ranking above a species and below a family, and composed of related species.

gills   Organs of aquatic animals (fish, insects, and others) for breathing in water.

gland   An organ in an animal's body that secretes a substance the animal needs, such as a salivary gland that secretes saliva.

gradual metamorphosis   See **metamorphosis.**

grub   The wormlike larva of many types of beetles.

habitat   The natural environment in which an animal or plant lives.

host   An animal or other living thing on which a parasite lives.

incomplete metamorphosis   See **metamorphosis.**

invertebrates   Animals without backbones.

kingdom   One of the five major divisions of living things (animals, plants, fungi, one-celled organisms, and blue-green algae).

larva (plural **larvae**)   Any immature insect that is very different in shape and lifestyle from the adult and that undergoes complete metamorphosis.

maggot   The headless, legless wormlike larva of many types of flies.

mammals   Animals that have backbones, are warm-blooded, and secrete milk for their young.

metamorphose   To undergo metamorphosis.

metamorphosis   A change in body form during development. A gradual change that occurs without a pupal stage is called **gradual** (or **incomplete) metamorphosis.** An abrupt and great change, as occurs in the cocoon of a moth, is called **complete metamorphosis.**

migration   The act of traveling each year from one region or climate to another for feeding or breeding.

mimicry   The resemblance of a living thing to some other living thing or object in the environment. Predators or prey may mistake the mimic for the model that it resembles. The deception improves the mimic's chances for survival or successful reproduction. In **Batesian mimicry,** the model has some feature that discourages predators (such as tasting bad), but the mimic does not. In **Müllerian mimicry,** the model and the mimic share an offensive trait.

molt   To shed the exoskeleton after a period of growth, as in insects.

Müllerian mimicry   See **mimicry.**

mutualistic relationship   A close relationship between two species that benefits both species.

nectar   A sugary substance produced by flowers to attract pollinators.

nocturnal   Active at night.

nutrient   A chemical substance in food that is needed for life.

nymph   The young stage of any insect species that undergoes gradual (or incomplete) metamorphosis.

order   A category of biological classification ranking above a family and below a class, and composed of related families.

parasite   An animal that feeds and is sometimes sheltered on another animal but does not kill it, at least not right away.

parental care   Care provided by a parent. In animals, this can include shelter, protection from predators, and food.

parthenogenesis   Reproduction by a female that has not mated. The offspring have no father and are identical to the mother.

pheromone   A chemical substance secreted by an animal that influences the behavior of others of the same species.

phylum (plural **phyla**)   One of the primary divisions of the animal kingdom, ranking above a class and below a kingdom, and composed of related classes.

pollen   Dustlike grains, usually yellow, that contain the sperm cells or male reproductive cells in flowering plants. An egg in a flower must be fertilized by the sperm cells in a pollen grain in order to produce a seed.

pollinator   An animal, usually an insect or hummingbird, that transports pollen from the male reproductive parts on one flower to the female reproductive parts on another flower. This usually happens as a result of the animal's flying from flower to flower in search of nectar.

predator   An animal that kills and eats other animals.

predator swamping   The emergence of prey in large numbers at one time, far exceeding the number that predators can eat. This ensures that many prey will survive.

prey   An animal hunted or caught for food.

proboscis   A tubular sucking or poking mouthpart, as in hemipterans or flies.

pronotum   The top covering of the prothorax of an insect. In some insects, it is large and conspicuous and provides the top cover of the body between the head and the base of the wings.

prothorax   The first of the three segments of the thorax of an insect, bearing the first pair of legs.

pupa (plural **pupae**)   In insects with complete metamorphosis, the stage of development where the larva changes to the adult form.

pupate   To undergo the process of changing from larva to adult.

queen   The fertile female of social insects, which is bigger than other individuals in the colony and whose only function is to lay eggs. The queen is fed and tended by workers.

replete   A worker ant whose crop is greatly enlarged with liquid food, so that the abdominal segments are pulled apart and the membranes in between are stretched tight. Repletes serve as living storage tanks, and regurgitate on demand to nest mates.

scavenger   An animal that eats dead plant or dead animal matter.

scutellum   A more or less triangular plate on the back and behind the pronotum of some insects. In stinkbugs, the scutellum does not extend as far down the back as the hard, thick part of the front wings.

social insects   Insects such as ants, bees, wasps, and termites that live in colonies with division of labor and complex social interactions.

soldier   A member of a colony of social insects that uses its large head and jaws to defend the colony by attacking and biting enemies. Soldiers are not able to reproduce.

species (plural **species)**   A category of biological classification ranking below a genus, and composed of living things capable of interbreeding.

spiracle   One of several openings along the sides of an insect's abdomen. Each opening leads into a breathing tube where an exchange of gases occurs.

swarm   A large group of moving insects.

symbiotic relationship   A close relationship between two species.

thorax   The body region behind the head of an insect to which the legs and wings are attached.

toxic   Harmful or deadly.

vertebrates   Animals having backbones, including fish, amphibians, reptiles, birds, and mammals.

warm-blooded   Having a body temperature that is regulated internally and stays constant.

worker   A member of a colony of social insects whose job is to hunt for food outside the nest and bring it back, feed and tend the queen and her eggs and larvae, make repairs on the nest, and so on. Workers are not able to reproduce.

Index